Harmony: Unveiling the Master Architect

Introduction

In Harmony we will explore compelling facts that point to an intelligent Creator behind the universe by bridging science, theology, and history. We present evidence that Jesus is the intelligent designer and creator of all, as John 1:3 tells us, "All things were made by him; and without him was not any thing made that was made." The theory of Intelligent Design proposes that certain features of the universe and living organisms are best explained by an intelligent cause rather than by natural selection.

The complex structure of DNA, with its precise coding, is a testament to a deliberate designer. Moreover, natural patterns—like the Fibonacci sequence evident in leaves, shells, and trees—reveal a mathematical order that points to a grand design instead of random chance. One example we explore in Harmony is what is referred to as "God's signature," a phenomenon in which our DNA appears to contain the name of the God of Abraham, YHVH.

By delving into the spiritual texts of the King James Bible, we aim to uncover hidden mysteries that enhance our understanding of an intelligent Creator and offer profound insights into the nature of existence and our relationship with the divine. For instance, we examine the first three words of Genesis 1:1, which, when studied alongside the corresponding Hebrew pictographs, seem to deliver a prophetic message about the Son of God meeting his destiny on the cross.

Another intriguing example is the design of the Israelite camp, constructed according to God's precise instructions around 1300 BC, which, when viewed from above, resembles a perfect cross.

We will also explore various non-biblical documents that reference Christ and early Christianity. Tacitus, in his Annals, mentions Christ (Christus) and his crucifixion under Pontius Pilate during Tiberius's reign, while The Antiquities of the Jews offers additional accounts of Christ and his crucifixion. Furthermore, we will investigate several prophecies that foretold the coming of Christ centuries in advance. For example, Psalm 22—composed about 1,000 years before

Christ's time—provides a remarkably accurate portrayal of the crucifixion in verses 16–18: "For dogs have compassed me: the assembly of the wicked have inclosed me: they pierced my hands and my feet. I may tell all my bones: they look and stare upon me. They part my garments among them and cast lots upon my vesture."

In addition, we will consider the historical reliability of the Bible, where archaeological discoveries and historical records validate many of its accounts. The consistent laws observed in physics, chemistry, and biology, which govern all matter and energy, further affirm the notion of a designed reality. From the predictable orbits of planets to the delicate balance found within ecosystems, the orderly structure of the universe suggests that an intelligent mind established these frameworks.

As we embark on this journey, let us contemplate Paul's words in Ephesians 5:14 "Wherefore he saith, Awake thou that sleepest, and arise from the dead, and Christ shall give thee light." It is essential to seek out the truth, acknowledge the nature of God, and understand our identity within Him. Created in God's image, our purpose is to live perpetually in love and gratitude, in perfect harmony with the divine—where evil does not exist, as assured by God in Hebrews 10:17: "Their sins and iniquities will I remember no more."

"It is the glory of God to conceal a thing: but the honour of kings to search out a matter."

Proverbs 25:2

Contents

Chapter 1. Intelligent Creator……………………..1

Chapter 2. Prophecies of Jesus…………………….42

Chapter 3. Jesus is God……………………………..107

Chapter 4. Historicity of Jesus……………………116

Chapter 5. Historicity of The Bible……………..124

Chapter 6. Science in The Bible…………………139

Chapter 7. 12 Universal Laws in The Bible……….148

Chapter 8. Salvation………………………………….156

Chapter 9. Psalms……………………………………176

Chapter 1

Intelligent Creator

The chapter begins by establishing the foundational concept of intelligent design through the examination of the Fibonacci sequence in nature. Starting with a mathematical explanation of the sequence (0, 1, 1, 2, 3, 5, 8, 13, 21...)

The chapter demonstrates how this divine pattern appears throughout creation. Specific examples include the DNA molecule's measurements (34 angstroms by 21 angstroms), flower petal arrangements (lilies with 3 petals, buttercups with 5, chicory with 21), spiral patterns in seashells and pinecones, tree branch formations, and even human body proportions.

The chapter extends this pattern to larger scales, showing how the Fibonacci spiral appears in hurricane formations and galaxy structures. This evidence is presented as a clear signature of intelligent design, arguing that such precise mathematical patterns throughout nature point to a divine Creator rather than random chance. Moreover, we will be exploring hidden mysteries in the Bible that enrich our understanding of Jesus as the creator of all.

Is God's Design Evident in the Fibonacci Sequence? (KJV)

In mathematics there are a variety of "special" number sequences: Prime Numbers, Odd Numbers (1,3 5,7, …), Even Numbers (2,4,6,8…); Rational and Irrational Numbers. But there is also a very special sequence of numbers called "The Fibonacci Sequence", after its discoverer, Fibonacci of course.

In mathematics, the Fibonacci numbers, commonly denoted F_n, form a sequence, the Fibonacci sequence, in which each number is the sum of the two preceding ones. The sequence commonly starts from 0 and 1, although some authors start the sequence from 1 and 1 or sometimes (as did Fibonacci) from 1 and 2. Starting from 0 and 1, the first few values in the sequence are:

0, 1, 1, 2, 3, 5, 8, 13, 21, 34, 55, 89, 144….

They are named after the Italian mathematician Leonardo of Pisa, later known as Fibonacci, who introduced the sequence to Western European mathematics in his 1202 book Liber Abaci.

Fibonacci numbers appear unexpectedly often in mathematics, so much so that there is an entire journal dedicated to their study, the Fibonacci Quarterly. They also appear quite often in biological settings, such as branching in trees, the arrangement of leaves on a stem, the fruit sprouts of a pineapple, the flowering of an artichoke, an uncurling fern, and the arrangement of a pinecone's bracts.

Fibonacci numbers are strongly related to the golden ratio: Binet's formula expresses the nth Fibonacci number in terms of n and the golden ratio and implies that the ratio of two consecutive Fibonacci numbers tends to the golden ratio as n increases. This shape, a rectangle in which the ratio of the sides a/b is equal to the golden mean (phi), can result in a nesting process that can be repeated into infinity — and which takes on the form of a spiral. This logarithmic spiral, or golden ratio, abounds in nature.

Phi and the Golden Ratio (or Rectangle)

As with other Greek letters, lowercase phi is used as a mathematical or scientific symbol. Some uses, such as the golden ratio, require the old-fashioned 'closed' glyph.

In mathematics, two quantities are in the golden ratio if their ratio is the same as the ratio of their sum to the larger of the two quantities. Mathematicians have studied the golden ratio's properties since antiquity. It is the ratio of a regular pentagon's diagonal to its side and thus appears in the construction of the dodecahedron and icosahedron. The golden ratio has been used to analyze the proportions of natural objects and artificial systems such as financial markets, in some cases based on dubious fits to data. The golden ratio appears in patterns in nature, including the spiral arrangement of leaves and other parts of vegetation.

Some 20th-century artists and architects, including Le Corbusier and Salvador Dalí, have proportioned their works to approximate the golden ratio, believing it to be aesthetically pleasing. These uses often appear in the form of a golden rectangle.

The Fibonacci Sequence is Clearly Shown in Nature

The Bible starts by proclaiming that an all-powerful God created ("bara" – to create out of nothing) the heavens and the earth.

"In the beginning God created the heavens and the earth (the cosmos)" Gen 1:1

That includes the sun, the moon, the stars, the galaxies, and of course the earth, with all its living and non-living things. Now if the all-powerful Creator and "Master Designer" was the Author of all creation, it shouldn't surprise us that we would see evidence of His design across all of nature. And indeed, we do. Evidence of the Fibonacci Sequence is spread across all of creation, from the design of DNA which resides in all living things, to seashells, plants and flowers, trees, and the design of human and animal bodies. The Fibonacci Sequence or "golden ratio" is also evident in the spiral nature of hurricanes, and even the spiral shape of galaxies! Here are just a few examples of where this "golden ration" shows up in nature:

The Golden Ratio Appears in the Shape of DNA molecules

Even the microscopic realm is not immune to Fibonacci. The DNA molecule measures 34 angstroms long by 21 angstroms wide for each full cycle of its double helix spiral.

These numbers, 34 and 21, are numbers in the Fibonacci series, and their ratio 1.6190476 closely approximates Phi, 1.6180339.

In Seeds and Flower petals

The number of petals in a flower consistently follows the Fibonacci sequence. Famous examples include the lily, which has three petals, buttercups, which have five, the chicory's 21, the daisy's 34, and so on. Phi appears in petals, since each petal is placed at 0.618034 per turn (out of a 360° circle) allowing for the best possible exposure to sunlight and other factors.

The head of a flower is also subject to Fibonaccian processes. Typically, seeds are produced at the center and then migrate towards the outside to fill all the space. Sunflowers provide a great example of these spiraling patterns. In some cases, the seed heads are so tightly packed that total number can get quite high — as many as 144 or more. And when counting these spirals, the total tends to match a Fibonacci number. Interestingly, a highly irrational number is required to optimize filling (namely one that will not be well represented by a fraction). Phi fits the bill rather nicely.

Many Seashells Display the Fibonacci Pattern

Snail shells and nautilus shells follow the logarithmic spiral, as does the cochlea of the inner ear. It can also be seen in the horns of certain goats, and the shape of certain spider's webs.

You See the Golden Ration in the Shape of Pinecones

Similarly, the seed pods on a pinecone are arranged in a spiral pattern. Each cone consists of a pair of spirals, each one spiraling upwards in opposing directions. The number of steps will almost always match a pair of consecutive Fibonacci numbers. For example, a 3-5 cone is a cone which meets at the back after three steps along the left spiral, and five steps along the right.

Fruits and Vegetables

Likewise, similar spiraling patterns can be found on pineapples and cauliflower.

The Fibonacci Sequence is in Tree Branches

The Fibonacci sequence can also be seen in the way tree branches form or split. A main trunk will grow until it produces a branch, which creates two growth points. Then, one of the new stems branches into two, while the other one lies dormant.

The Fibonacci Sequence is clearly evident in the way the tree sprouts as it grows – from a single trunk to two great limbs, then three, then five, then eight, and so on. This pattern of branching is repeated for each of the new stems. A good example is the sneezewort. Root systems and even algae exhibit this pattern.

Hurricanes Display the Fibonacci Spiral

The unique Fibonacci spiral is also evident in the pattern of hurricanes. Notice that the Bible mentions how God controls those clouds, and makes that lightening flash:

Job 37 14-16 Hearken unto this, O Job: stand still, and consider the wondrous works of God.

[15] Dost thou know when God disposed them, and caused the light of his cloud to shine?

[16] Dost thou know the balancings of the clouds, the wondrous works of him which is perfect in knowledge?

Spiral Galaxies Also Follow the Familiar Fibonacci pattern

Not surprisingly, spiral galaxies also follow the familiar Fibonacci pattern. The Milky Way has several spiral arms, each of them a logarithmic spiral of about 12 degrees. As an interesting aside, spiral galaxies appear to defy Newtonian physics. As early as 1925, astronomers realized that, since the angular speed of rotation of the galactic disk varies with distance from the center, the radial arms should become curved as galaxies rotate. Subsequently, after a few rotations, spiral arms should start to wind around a galaxy. But they don't — hence the so-called winding problem. The stars on the outside, it would seem, move at a velocity higher than expected — a unique trait of the cosmos that helps preserve its shape. Is it any wonder that the Bible declares:

Psalm 19:1 The heavens declare the glory of God; and the firmament sheweth his handywork.

Bodies, Faces and even Your Fingers Display Its Design

Although every person's body is different, proportional averages across populations tend to follow the Fibonacci Sequence. It has also been said that the more closely our proportions adhere to phi, the more "attractive" those traits are perceived. As an example, the most "beautiful" smiles are those in which central incisors are 1.618 wider than the lateral incisors, which are 1.618 wider than canines, and so on. It's quite possible that we are primed to like physical forms that adhere to the golden ratio — a potential indicator of reproductive fitness and health.

Faces abound with examples of the Golden Ratio. The mouth and nose are each positioned at golden sections of the distance between the eyes and the bottom of the chin. Similar proportions can be seen from the side, and even the eye and ear itself (which follows along a spiral). Photo by Andrea Piacquadio, Pexels.

Psalm 139:14 I will praise thee; for I am fearfully and wonderfully made: marvellous are thy works; and that my soul knoweth right well.

It's in the very design of your fingers! Looking at the length of our fingers, each section — from the tip of the base to the wrist — is larger than the preceding one by roughly the value of the golden ratio. In the Way Life Reproduces

One also sees the Fibonacci Sequence in the way life reproduces. For example, the honeybee follows Fibonacci in interesting ways. The most profound example is by dividing the number of females in a colony by the number of males (females always outnumber males). The answer is typically something very close to 1.618. In addition, the family tree of honeybees also follows the familiar pattern. Males have one parent (a female), whereas females have two (a female and male). Thus, when it comes to the family tree, males have 2, 3, 5, and 8 grandparents, great-grandparents, gr-gr-grandparents, and gr-gr-gr-grandparents respectively. Following the same pattern, females have 2, 3, 5, 8, 13, and so on. And as noted, bee physiology also follows along the Golden Curve rather nicely.

It's Evident Even in the Flight Patterns of Birds

When a hawk approaches its prey, its sharpest view is at an angle to their direction of flight — an angle that's the same as the spiral's pitch

Job 12:7-10 But ask now the beasts, and they shall teach thee; and the fowls of the air, and they shall tell thee:

[8] Or speak to the earth, and it shall teach thee: and the fishes of the sea shall declare unto thee.

[9] Who knoweth not in all these that the hand of the Lord hath wrought this?

[10] In whose hand is the soul of every living thing, and the breath of all mankind.

God's Workmanship is Clearly Seen in Nature!

God speaks to us through general revelation His creation, and well as through His special revelation, the Bible. When you look up and behold the sun, moon and stars, He speaks to you about His divine nature and power:

The Bible emphasizes that God has made His divine nature and godhead plain, by the things which are clearly seen:

Rom 1:19-20 Because that which may be known of God is manifest in them; for God hath shewed it unto them.

[20] For the invisible things of him from the creation of the world are clearly seen, being understood by the things that are made, even his eternal power and Godhead; so that they are without excuse.

The Fibonacci Pattern is Nature: By Accident, or Evidence of Design?

The fact is that the Fibonacci Sequence shows up in nature and is clearly "seen" – from the smallest life forms to entire galaxies.

There are two views about such a pattern of numbers "showing up" in nature. Those that believe in evolution, the notion that all things that exist today in nature were produced by natural selection acting on random mutations in organisms over a long period of time, will tell you that you are only seeing the "appearance of design". That there really is no intentional design, or Designer. It's all just a product of blind chance.

The other view is that there really are patterns of design in nature, and that they reflect the handiwork of an all-powerful Designer acting through creative processes to produce this design pattern across all of nature – from the smallest plants and animals to the design of entre galaxies. When you look at a beautiful lace curtain, you picture a curtain designer. An intricate watch speaks of an ingenious watchmaker. It's patently obvious that where there is a design, there is a Designer. These things just don't happen by accident.

Which is a more reasonable conclusion, based on the evidence? These patterns all across nature are the product of blind chance evolution, or the product of an all-powerful, all-knowing, Creator Designer working through the natural processes He started and continues to work through? You decide.

How Would You Like to Meet the Original Author and Designer?

John 3:16 For God so loved the world, that he gave his only begotten Son, that whosoever believeth in him should not perish, but have everlasting life.

Sources: Evidence To Believe in: The Fibonacci Sequence – God's Design Pattern is Everywhere, Wikipedia: Fibonacci Numbers, Phi, and the Golden Ratio, John Kostik: The Divine Proportion, George Dvorsky: 15 Uncanny Examples of the Golden Ratio in Nature.

GOD'S SIGNATURE

IN EVERY CELL OF THE HUMAN BODY

DNA
DEOXYRIBO NUCLEIC ACID

4 Nuclide Acids bind the Helixes together
by Sulfuric Bridges in the sequence of

A-T-C-G

EVERY 10, 5, 6, 5 ACIDS

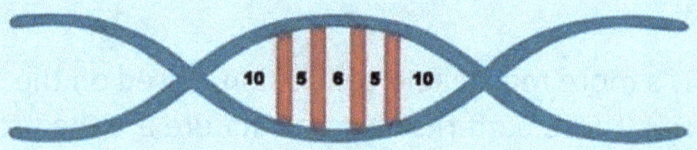

HEBRAIC NUMERIC MEANING

10   5   6   5
Y-H-V-H

The Bridge, Bond, Glue that keeps the Human DNA Sequence together.

COLOSSIANS 1:16-17
For in Him [JESUS] all things were created, things in Heaven and on Earth,
visible and invisible, whether Thrones or Dominions or Rulers or Authorities.
All things were created through Him and for Him.
He is before all things, and in Him
ALL THINGS HOLD TOGETHER.

Yeshua and the Hebrew Alphabet
The Aleph and the Tav (KJV)

Yeshua the Messiah testified that He is the *Aleph* and the *Tav,* the First *(rishon)* and the Last *(acharon),* and the Beginning *(rosh)* and the Ending *(sof)*:

אָנֹכִי אָלֶף וְתָו רִאשׁוֹן וְאַחֲרוֹן רֹאשׁ וָסוֹף

Rev. 22:13 I am Alpha and Omega, the beginning and the end, the first and the last.

When Yeshua said this, he was making a direct reference to Isaiah 41:4, 44:6, and 48:12, where Adonai Himself says that He is the First and the Last -- and explicitly declared that there is no other "god" beside Him.

Please get a hold of the implication here: Jesus of Nazareth was claiming that He was the one to whom the references in Isaiah pertain. He is the "Direct Object" of the universe - the divine Agent of the Scriptures (see below about the role of the direct object marker).

אָנֹכִי הַדֶּרֶךְ וְהָאֱמֶת וְהַחַיִּים
וְאִישׁ לֹא־יָבֹא אֶל־הָאָב בִּלְתִּי עַל־יָדִי

Yeshua is the Truth...

Yeshua also said He was the Truth of God Himself

John 14:6 *Jesus saith unto him, I am the way, the truth, and the life: no man cometh unto the Father, but by me.*

Notice that the Hebrew word for truth (i.e., *emet:* אֱמֶת) contains the first letter Aleph (א), the middle letter Mem (מ), and the last letter Tav (ת) of the Hebrew alphabet, which the Jewish sages say implies that the truth contains everything from Aleph to Tav:

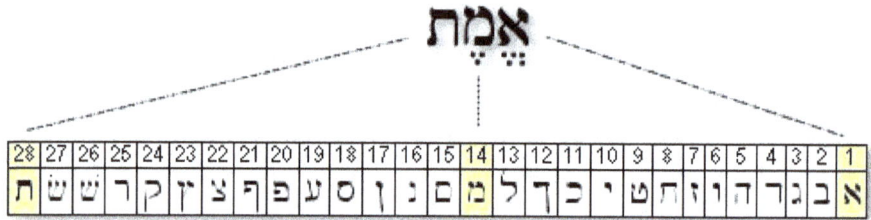

What is Truth?

The Hebrew word *emet* has a more concrete meaning than the English word for "truth" (the English word derives from the Greek/Western view of truth as a form of correspondence between language and reality, but invariably languished over epistemological questions that led, ultimately, to skepticism). In the Hebraic mindset, the person who acts in *emet* is one who can be *trusted* (Gen. 24:49; 42:16; 47:26; Josh. 2:14). Actions, speech, reports, or judgment are *emet* because they are *reliable* (Deut. 13:14; 22:20; 1 Kings 10:6; 22:16; Prov. 12:19; Zech. 8:16). If a seed is a seed of *emet*, its quality is *trustworthy* (Jer. 2:21).

In the Tanakh, *emet* is often coupled with *chesed,* covenant faithfulness, which designates God's loyalty in fulfilling his promises and his covenant. For example, God's *emet* and *chesed* were majestically revealed in giving the covenant at Sinai (Ex. 34:6).

יְהוָה יְהוָה אֵל רַחוּם וְחַנּוּן אֶרֶךְ אַפַּיִם וְרַב־חֶסֶד וֶאֱמֶת

Exodus 34:6 And the Lord passed by before him, and proclaimed, The Lord, The Lord God, merciful and gracious, longsuffering, and abundant in goodness and truth,

Indeed, Pilate's question, "What is truth?" is a *category* mistake, since truth is not about "what" but about "Who." That is, truth is not something objective and static, a thing to be known and studied from a distance. No. Truth is essentially personal. It is personal disclosure of the character of the subject. Understood in this way, truth is a way of living, a mode of existence, a relational truth.

הוּא אוֹר אֱמֶת אֲשֶׁר בָּא לָעוֹלָם לְהָאִיר לְכָל־אָדָם

John 1:9 That was the true Light, which lighteth every man that cometh into the world.

Yeshua the Divine Direct Object...

Interestingly, Aleph and Tav form a unique word that functions as a "direct object marker" in both Biblical and modern Hebrew:

As it is written in Genesis 1:1, In the beginning God created the heaven and the earth.

בְּרֵאשִׁית בָּרָא אֱלֹהִים אֵת הַשָּׁמַיִם וְאֵת הָאָרֶץ

Considered this way, Jesus is the Direct Object of the Universe, the End *(sof)* of all of creation. And not only is Jesus the End of all creation, but He is the "Beginning of the Creation of God," the Creator and Sustainer of all things:

Colossians 1:16-17 For by him were all things created, that are in heaven, and that are in earth, visible and invisible, whether they be thrones, or dominions, or principalities, or powers: all things were created by him, and for him:

[17] And he is before all things, and by him all things consist.

כֹּה אָמַר הָאָמֵן עֵד הָאֱמֶת וְהַצֶּדֶק וְרֵאשִׁית בְּרִיאַת הָאֱלֹהִים

Rev. 3:14 And unto the angel of the church of the Laodiceans write; These things saith the Amen, the faithful and true witness, the beginning of the creation of God;

Yeshua is the Strong Sign

Finally, using the ancient pictographs, we can see that Jesus is the "Strong Sign" from Adonai:

He is the One who comes in humble, silent strength (Aleph) bearing the Sign of the true Covenant of God (Tav). *Source: John Parsons, Hebrew 4 Christians.*

Biblical "Hebrew to English" Alphabet
Hebrew is the first and oldest alphabet: 1859 BC

Pictograms		Phonograms		Echograms	Aramaic Hebrew	Masoretic Hebrew
Egyptian Hieroglyphics 1859 BC		Mosaic Hieroglypic Hebrew Alphabet 1859 - 550 BC		English Modern	Square Hebrew 550 BC - 70 AD	Vowelled Hebrew 600 AD - present
Gardiner's Sign List #	Sounds Like	First Hebrew Phonogram Alphabet 1859 - 1100 BC	Paleo-Hebrew 1100-550BC	English	First Century	Vowels, dots, dashes were invented by Masoretes (600 AD) did not exist before.
(F1)	K	Aleph Cattle		A	א	Silent stop, like the "-" in "A-ha".
(O1) (O4)	Pr /H	Bayit House		B,V	ב	ב B as in Bet (With dot) / ב V as in Vet
(O38)	Knbt	Gahar Bend		G	ג	G as in Gift
(O31)	.	Delet Door		D	ד	D as in Door
(A28)	Hi	Halal Praise		H,E	ה	H as in Hay
(O30)	Shnt	Vaw Pillar Support		V,O,U	ו	V as in Vine / Vowel "u" as in "Flute" / Vowel "o" as in "Hole"
(D13)	inh	Zeah Sweat (Brows)		Z	ז	Z as in Zechariah
(O6) (V28)	Hwt /H	Haser/Hut Enclosure/Thread		H,Ch	ח	Ch as in Bach
(F35)	D	Tov Good		Th	ט	Th as in Thin
(D36) (D47)	A	Yad Hand		I,Y,J	י	Y as in Yes / Vowel "i" as in machine / Vowel "ey" as in "they".
(D28)	K	Kap Palms		K,Ch	כ,ך	כ K as in King (With dot) / כ Ch as in Bach
(S39)	Wt	Lamad Teach		L	ל	L as in Learn
(N35)	N	Mayim Water		M	מ,ם	M as in Memory
(I9) (I10)	F	Nahas Snake		N	נ,ן	N as in Now
(D3) (K5)	./Bz	Sear/Sarah Hair/Stink		S,X	ס	S as in Support
	IR	Ayin Eye		O	ע	Silent guttural in the back of the throat
D21	R	Peh Mouth		P,Ph	פ,ף	פ P as in Power (with dot) / פ Ph as in Phone
V33	Ssr	Seror Sack		Ts	צ,ץ	Ts as in Sits
V25	Wd	Qur Spun fiber		Q	ק	C as in Cry (more guttural than Kaph)
D1	Tp	Resh Head		R	ר	R as in Rush
D27	Mnd	Sadayim Breasts		S,Sh	ש	ש Sh as in Shine (right dot) / ש S as in Sun (left dot)
M42	Wn	Tayis Male goat		T,Th	ת	ת T as in Time (dot) / ת Th ans in Theme

As we see here using ancient Hebrew alphabet and pictographs, Genesis 1:1 foretells the coming of Jesus,

(The) Son (of) God (would be) Destroyed (by his own) Hand (on the) Cross.

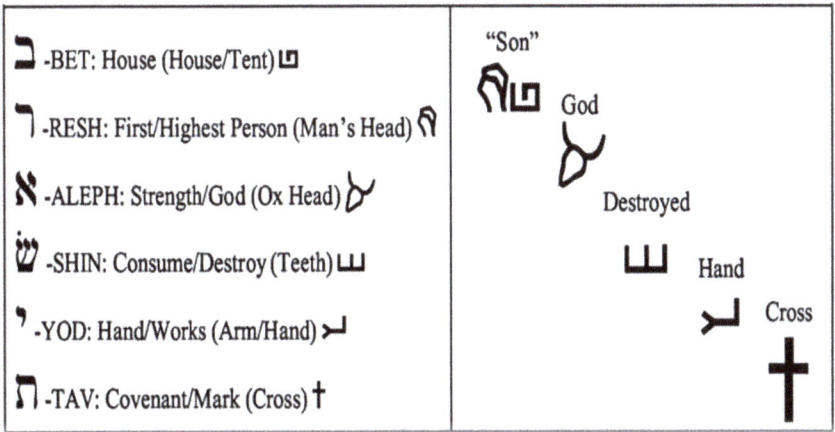

Name Meanings from Adam to Noah

Hebrew	English
Adam	Man
Seth	Appointed
Enosh	Mortal
Kenan	Sorrow
Mahalelel	The Blessed God
Jared	Shall come down
Enoch	Teaching His death shall bring
Methuselah	The despairing
Noah	Rest or comfort

The Camp of Israel

The great discovery which opens the Bible to us is that these 66 books, written by 40 authors over thousands of years, are an integrated message system. Every detail--every number, every place name, every allusion--is there by careful design and is significant.

Jesus Himself highlighted this in Matthew 5:17-18

Think not that I am come to destroy the law, or the prophets: I am not come to destroy, but to fulfill. For verily I say unto you, Till heaven and earth pass, one jot or one tittle shall in no wise pass from the law, till all be fulfilled.

This insight unlocks a number of discoveries which otherwise would escape the notice of someone who takes the text less seriously.

What insights are hidden in the numbering of the Tribes of Israel?

In this brief article we'll review a remarkable perspective from a passage that most of us might skip over.

Numbering the People

In Numbers chapter 1 we encounter the numbering of the people. Why? Why did the Holy Spirit want you to know this list of numbers? What hidden insight lies behind them?

Of course, there are valid historical reasons for the inclusion of this detail in the Torah (the five books of Moses). But if we examine these details more closely, some remarkable insights emerge.

The Tabernacle

When Moses received the Ten Commandments on Mt. Sinai, he also received detailed specifications and instructions for the building of a portable sanctuary, the Tabernacle, or tent of meeting. (1) The purpose of this unusual facility was to provide a place for God to dwell among His people. (A review of the specifies of this remarkable structure will be reserved for a subsequent article.)

The tabernacle was always set up at the center of the Camp of Israel. The tribe of Levi was assigned to care for it and encamped around it. Moses, Aaron, and the priests camped on the east side next to the entrance. The three families of the tribe of Levi (Merari, Kohath, and Gershon), camped on the north, south and west side, respectively.

The remaining twelve tribes were grouped into four camps around the Levites.

A Baker's Dozen

It is helpful to realize that there were really 13 tribes, not just "twelve." This can be confusing to the uninitiated reader.

Jacob had twelve sons, each becoming the founder of one of the twelve tribes. However, Joseph was sold into slavery and subsequently emerged as the prime minister of Egypt. (2) in Egypt, Joseph married Asenath and had two sons, Manasseh and Ephraim. When Jacob and the rest of the family ultimately came to Egypt, Jacob adopted his two grandsons as his own. (3) With the tribe of Joseph in two parts, we have an "alphabet" of 13 to choose from.

The twelve tribes of Israel (Jacob) are listed twenty times in the Old Testament. (4) They are listed by mother (Leah, Rachel, Zilhah, and Bilhah), their numeration, their encampment, order of march, their geographical relations, etc. Each time they are listed in a different order.

The Levites were exempt from military duties. When the order of military march is given, there are still 12 listed, excluding Levi. How? By dividing Joseph into two: Ephraim and Manasseh.

(Levi is thus omitted on four occasions. In a similar manner, Dan is omitted on three occasions, the most notable one in Revelation 7.)

The Four "Camps"

The twelve remaining tribes, excluding the Levites, were clustered into four "camps." (5) Each of these groups, of three tribes each, were to rally around the tribal standard of the lead tribe.

Judah's tribal standard was, of course, the lion. Reuben's ensign was a man; Ephraim's the ox; Dan's, ultimately, the eagle. These are detailed in the following diagram.

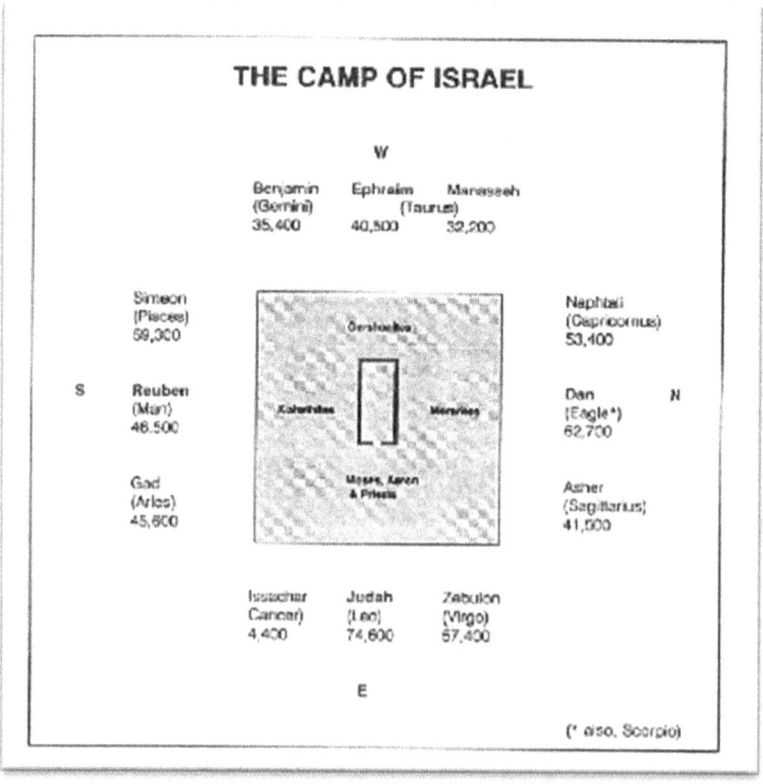

The Mazzeroth

It may come as a surprise to many to learn that each of the 12 tribes were associated with one of the constellations of the mazzeroth (the Hebrew zodiac.) We know these by their post-Babel names after being corrupted by pagan traditions. By learning the Hebrew names, and the names of the principal stars in the order of their magnitude, we discover they portray the entire redemptive plan of God--from the virgin birth (Virgo) to the triumph of the Lion of the Tribe of Judah (Leo).

The Four Faces

It is interesting to note that these four primary tribal standards--the lion, the man, the ox, and the eagle--are the same as the four faces of the cherubim. Each time we encounter a view of the throne of God, (6) we notice these strange living creatures, somehow associated with the protection of His throne, His holiness, etc.

It would seem that the camp of Israel--with the tabernacle in the middle--seems to be a model of the throne of God: His presence in the center, represented by the tabernacle, encircled by the four faces, all surrounded by His people.

But there's even more. Why the specific numbers?

The Numbering

The numbering of the tribes is detailed in Numbers chapter

1. The actual population represented is obviously somewhat larger than these enumerations, since only men over twenty, able to go to war, were counted. Most analysts assume that women, children, and the elderly, would multiply the count by some factor: 3 or whatever. The total camp would thus appear to approximate two million.

While the numbers of each tribe may not seem to be very revealing, the totals for each of the four camps are.

Cardinal Compass Points

Each of the camps, of three tribes each, were to encamp on one of the cardinal compass directions (N, S, E, or W) with respect to the camp of the Levites enclosing the tabernacle. (7)

We can only guess at how much space was required by the Levites, whether it was 100 ft. on a side, 100 yards, or whatever. But whatever it was, we'll view that length as a basic unit.

To fully appreciate all of the implications, you must try to think like a rabbi: you need to maintain an extremely high respect for the precise details of the instructions.

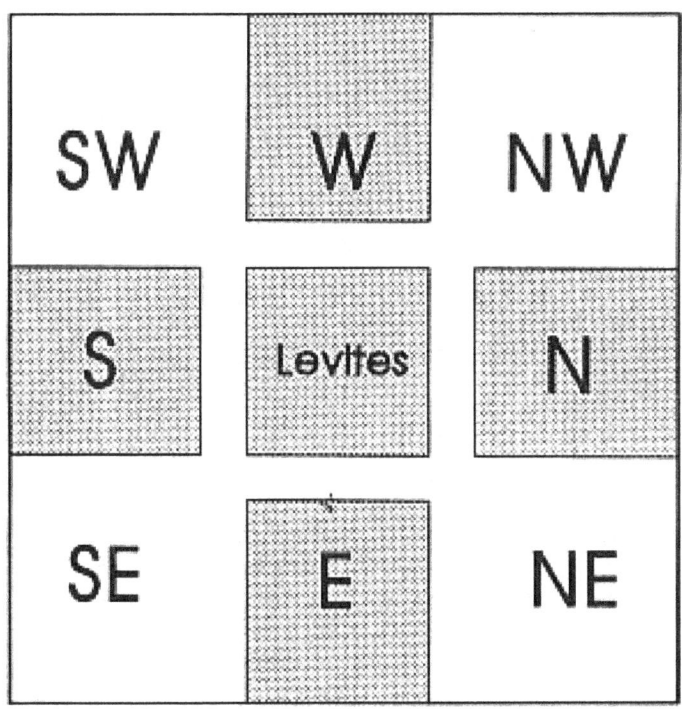

The tribes of Judah, Issachar, and Zebulun--collectively called the Camp of Judah--had to encamp east of the Levites. This poses a technical problem. Notice that if the breadth of their camp was larger than that of the Levites, the excess would be southeast or northeast, not east. Therefore, their camp could only be as wide as the Levites, and they then had to extend eastward to obtain whatever space they required.

The camps of Reuben, Ephraim, and Dan had the same constraint on the south, west, and north, respectively. The length of each leg would be proportional to the total in each camp.

Aerial View

If we assemble what we can infer from the Torah account, we can imagine what the camp of Israel looked like from above: the tabernacle and the Levites in the center, surrounded by the four faces of the tribal standards, and each of the four camps of Judah, Ephraim, Reuben, and Dan, stretching out in the four cardinal directions.

We can also tally the size of each tribe to total the relative length of each camp as they stretched out in each of the four directions. The plan view, on a relative scale, is shown below.

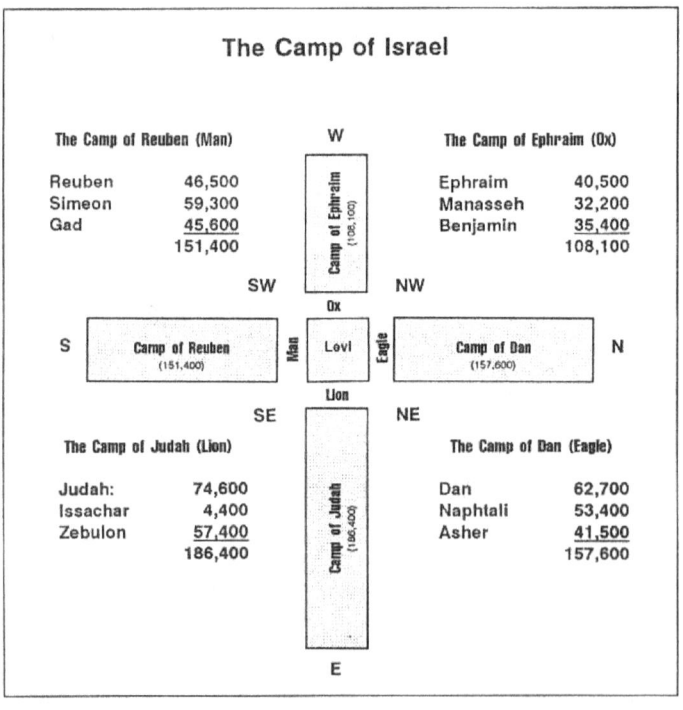

It would appear to us that it is a cross! Isn't that remarkable? And this is from the Torah, not the New Testament!

The New Testament is in the Old Testament concealed; The Old Testament is in the New Testament revealed.

Source: Chuck Missler Koinonia House Ministries.

Abraham and Issac: The Akedah

Perhaps the most startling example of a "type" is the famed incident of Abraham offering his son Issac in Genesis Chapter 22, called in Hebrew the Akedah.

2] And [God] said, take now thy son, thine only son Isaac, whom thou lovest, and get thee into the land of Moriah;

and offer him therefore a burnt offering upon one of the mountains which I will tell thee of.

This is a strange call. Does God hereby endorse child sacrifice? Hardly! But then, what is going on? This episode has confused some scholars for centuries.

3] And Abraham rose up early in the morning, and saddled his ass, and took two of his young men with him, and Isaac his son, and clave the wood for the burnt offering, and rose up, and went unto the place of which God had told him.

By the time Abraham gets to Genesis 22, he has learned many lessons. Notice that he doesn't dally; he starts on his journey the very next morning!

Notice also that there are four going on the trip: Abraham, Issac, and two young men, as well as the donkey.

[4] Then on the third day Abraham lifted up his eyes and saw the place far off.

[5] And Abraham said unto his young men, Abide ye here with the ass; and I and the lad will go yonder and worship, and come again to you.

It took three days to get to the place now known as Mount Moriah. Notice also that the two young men remain at the base of the hill as the father and son climb up it.

Mount Moriah is a ridge system between the Mount of Olives to the east and Mount Zion to the west. It is bounded by the Kidron Valley on the east, the Tyropean Valley on the west, and the Hinom Valley to the south.

The ridge begins at the south at about 600 metres above sea level and rises to a peak as one goes northward. At the base of this ridge was the town of Salem at which Melchizedek was both the king and the priest. This later becomes Ophel, the city of David, and ultimately, Jerusalem.

Higher on the ridge, at about 741 metres above sea level is a saddle point where Ornan later owned a threshing floor, which would later be purchased by David to become the site of Solomons Temple.

(A threshing floor was not necessarily at the peak; it was typically a saddle point which enjoyed a prevailing wind which could be used to separate the chaff from the grain when threshed at harvest times.) The peak of the Mount is a bit further north, at about 777 metres above sea level, at a place which would later become known as Golgotha-the exact spot where Jesus Christ would be crucified as the offering for sin 2,000 years later.

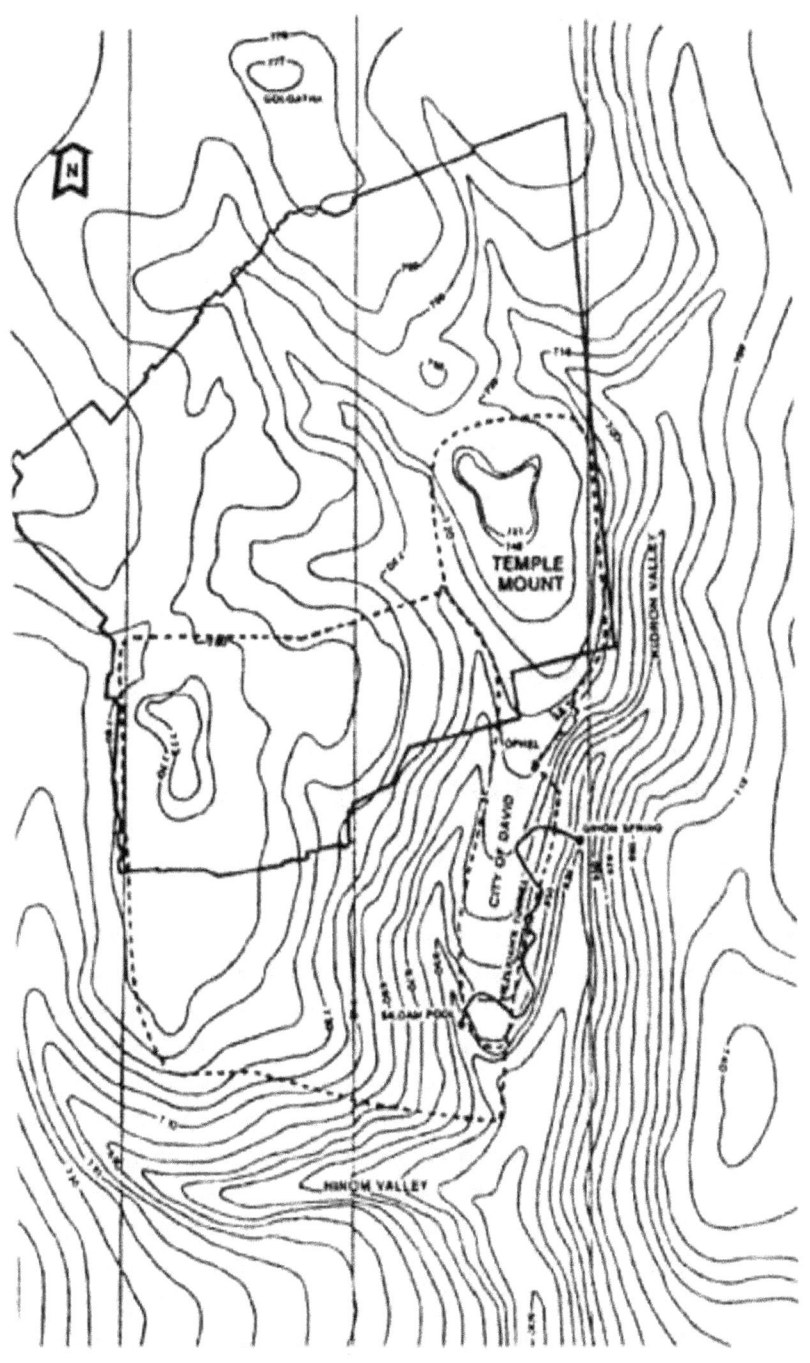

As we begin to understand the typology of this narrative, we begin to appreciate the subtleties in the account. Abraham had an earlier son, Ishmael, but for God's purposes here, Issac is viewed as "your only son"

Careful students of the Scripture have noted the law of first mention: that the first occurrence of a word in the Scripture is usually a very significant instance in the overall design. It is profoundly significant that this account includes the first occurrence of the word love in the Scripture.

[6] And Abraham took the wood of the burnt offering and laid it upon Issac his son; and he took the fire in his hand, and a knife; and they went both of them together.

[7] And Issac spake unto Abraham his father, and said, My father: and he said, Here am I, my son. And he said, Behold the fire and the wood: but where is the lamb for a burnt offering?

Good question, Issac. You can appreciate the lads concern. Notice Abrahams response:

[8] And Abraham said, My son, God will provide himself a lamb for a burnt offering: so they went both of them together.

"God will provide" who? Himself!

Was this also just a stall? Did Abraham realize that he was acting out a prophecy? Two thousand years later—on that very spot—another Father would offer His Son as the offering of all time!

And they came to the place which God had told him of; and Abraham built an altar there, and laid the wood in order, and bound Isaac his son, and laid him on the altar upon the wood.

We are all victims of our Sunday school coloring books: we always picture Isaac as a small boy. Some scholars maintain that Isaac was about 30 years old.

10] And Abraham stretched forth his hand, and took the knife to slay his son. 11] And the angel of the LORD called unto him out of heaven, and said, Abraham, Abraham: and he said, Here am I. 12] And he said, Lay not thine hand upon the lad, neither do thou any thing unto him: for now I know that thou fearest God, seeing thou hast not withheld thy son, thine only son from me. 13] And Abraham lifted up his eyes, and looked, and behold behind him a ram caught in a thicket by his horns: and Abraham went and took the ram and offered him up for a burnt offering in the stead of his son.

Thus, we encounter the substitutionary ram. When Adam and Eve "fell" in the Garden of Eden, even then, God taught them that by the shedding of innocent blood they would be covered. All the Levitical sacrifices in the Torah were designed to anticipate the climactic sacrifice for all time, also foreshadowed here. We are the beneficiaries of a love story, written in blood on a wooden cross, which was to be erected in Judea some two thousand years later.

14] And Abraham called the name of that place Jehovah-jireh: as it is said to this day, In the mount of the LORD it shall be seen.

Abraham then gave the location a prophetic label. It appears that he somehow knew that he was acting out a prophecy! Abraham also knew that Isaac, if offered, would have to be resurrected since God had previously promised Abraham that Isaac would have numerous descendants. (It is interesting that Isaac was "dead" to Abraham for three days: from the time the commandment came until he was freed by the angel.) Here, in this "type," or macrocode, we have Abraham cast in the role of the Father; and Isaac as the Son. The ultimate drama of God the Father, offering His Son, are the referents or designate to which this historical narrative appears to be alluding. There is even another subsequent example in which, again, the roles, or referents, are the same.

Weissmandl's Strange Legacy

(The following was first revealed to me by my friend, Dr. Gerald Schroeder, the famed atomic physicist who presently resides in Jerusalem. 24)

Here are the opening verses in the Book of Genesis in Hebrew: (Remember, Hebrew goes from right to left!)

בְּ־רֵאשִׁית בָּרָא אֱלֹהִים אֵת הַשָּׁמַיִם וְאֵת הָאָֽרֶץ׃
וְהָאָרֶץ הָיְתָה תֹהוּ וָבֹהוּ וְחֹשֶׁךְ עַל־פְּנֵי תְהוֹם וְרוּחַ
אֱלֹהִים מְרַחֶפֶת עַל־פְּנֵי הַמָּֽיִם׃
וַיֹּאמֶר אֱלֹהִים יְהִי אוֹר וַֽיְהִי־אֽוֹר׃
וַיַּרְא אֱלֹהִים אֶת־הָאוֹר כִּי־טוֹב וַיַּבְדֵּל אֱלֹהִים בֵּין הָאוֹר
וּבֵין הַחֹֽשֶׁךְ׃
וַיִּקְרָא אֱלֹהִים ׀ לָאוֹר יוֹם וְלַחֹשֶׁךְ קָרָא לָיְלָה וַֽיְהִי־עֶרֶב
וַֽיְהִי־בֹקֶר יוֹם אֶחָֽד׃ פ

The word Torah, in Hebrew, is four letters, . If you go to the first (tau, which is similar to our "T"), and count an interval of 49 letters, the next letter is a (vav, operating here like an "O"); count another interval of 49 letters and you will find a (resh, like our "R"); and then count another interval of 49 letters and you will find a (heh, or "H"). We find the word , or Torah, spelled out with 49 letter intervals. Rather strange. It would seem that someone has gone to some remarkable effort; and yet some argue that it is just coincidence. And when we examine the next book, the Book of Exodus, we discover the same thing again! Here are the first few verses of Exodus:

אֵלֶּה שְׁמוֹת בְּנֵי יִשְׂרָאֵל הַבָּאִים מִצְרָיְמָה אֵת יַעֲקֹב אִישׁ וּבֵיתוֹ בָּאוּ:
רְאוּבֵן שִׁמְעוֹן לֵוִי וִיהוּדָה:
יִשָּׂשכָר זְבוּלֻן וּבִנְיָמִן:
דָּן וְנַפְתָּלִי גָּד וְאָשֵׁר:
וַיְהִי כָּל־נֶפֶשׁ יֹצְאֵי יֶרֶךְ־יַעֲקֹב שִׁבְעִים נָפֶשׁ וְיוֹסֵף הָיָה בְמִצְרָיִם:
וַיָּמָת יוֹסֵף וְכָל־אֶחָיו וְכֹל הַדּוֹר הַהוּא:

Could this also be a coincidence, again? Just what are the chances of such a coincidence? The word might, on merely a statistical basis, appear in Genesis quite a few times depending on the range of intervals chosen. The total number of letters in Genesis is 78,064, and the amount of the letters, 4152; , 8448; , 4793, and, 6283. Indeed, appears three times in Genesis at the interval of 50, which is what might be statistically expected from a book of that length and of similar concentration of these four letters. But there is no reason why these should begin with the first of the book, and why this should happen in both Genesis and Exodus. The probability of such a coincidence has been estimated at about one in three million!

In the next book, the Book of Leviticus, this 49-letter interval doesn't seem to appear. (We'll return to reexamine an alternative discovery.)

When we examine the next book, the Book of Numbers, we discover that it happens again if we spell Torah backwards!

וַיְדַבֵּר יְהוָה אֶל־מֹשֶׁה בְּמִדְבַּר סִינַי בְּאֹהֶל מוֹעֵד בְּאֶחָד
לַחֹדֶשׁ הַשֵּׁנִי בַּשָּׁנָה הַשֵּׁנִית לְצֵאתָם מֵאֶרֶץ מִצְרַיִם לֵאמֹר:
שְׂאוּ אֶת־רֹאשׁ כָּל־עֲדַת בְּנֵי־יִשְׂרָאֵל לְמִשְׁפְּחֹתָם לְבֵית
אֲבֹתָם בְּמִסְפַּר שֵׁמוֹת כָּל־זָכָר לְגֻלְגְּלֹתָם:
מִבֶּן עֶשְׂרִים שָׁנָה וָמַעְלָה כָּל־יֹצֵא צָבָא בְּיִשְׂרָאֵל תִּפְקְדוּ
אֹתָם לְצִבְאֹתָם אַתָּה וְאַהֲרֹן:

When we examine the final book of the Torah, the Book of Deuteronomy, a similar thing happens, but again, backwards! Laying out the overall pattern:

Genesis   Exodus   Leviticus   Numbers   Deuteronomy
TORH →   TORH →   ?   ← HROT   ←   HROT

This seems to be too deliberate to be ascribed simply to chance. But why has this ostensibly deliberate arrangement been composed? What are the implications?

When we return to reexamine the Book of Leviticus, we discover that the square root of 49, 7, yields a provocative result. After the first yod, and an interval of seven, taking the next letter yields, the tetragramaton, the ineffable name of God, the YHWH. It appears that the Torah always points toward the Ineffable Name of God!

Genesis   Exodus   Leviticus   Numbers   Deuteronomy
TORH →   TORH →   YHWH   ← HROT   ←   HROT

This seems to hint of a hidden signature. Just as certain authors adopted a trademark, or "shtick," such as Alfred Hitchcock always appearing as an extra in his famous movies, or J.M.W. Turner's secret signature on his venerated watercolors, or the fabled hidden signature of Shakespeare in Psalm 46, we detect here evidence of hidden but deliberate design. And it may be a signpost pointing to others.

As we discovered in the previous chapter, within the first of the five books of Moses, known in Hebrew as the Torah, God's redemptive program was anticipated in the hidden message in the genealogy of Noah. Even the very name of this most venerated part of the Old Testament, highlights God's program.,

Even the word "Torah" itself, drawing on the concepts that lie behind the original Hebrew letters, embodies some provocative elements: The Tav (originally, a cross), the Vav (a nail), the Resh (the head of a man), and the Heh, (the breath or Spirit of God). Thus, Man, with the Spirit of God, nailed on the Cross! This term was in existence well before Messiah walked on the earth.

It is an interesting summary of the climax of God's love story, which was nailed on a cross erected in Judea 2,000 years ago. The entire Biblical drama records the extremes our Creator has resorted to in order to redeem man—including you and me—from our predicament.

Could this hidden design be simply an accident? There are those that argue that this is all a result of random chance. There are others who simply ascribe this remarkable structure to some ancient "diddling" by a clever scribe. However, we will discover that this all appears to be part of an even larger design. This discovery by Weissmandl appears to be only a remez, a hint of something hidden or something deeper.

It would be Weissmandl's rediscoveries of the ancient sages that subsequently inspired the Israeli researchers, 60 years later and armed with computers, to uncover what has now erupted into the modern day controversies, codes which appear to describe events which transcend the time period they were written: the events surrounding the revolt of the Maccabees; the storming of the Bastille during the French revolution; the holocaust in Germany; the treatment of diabetes; a description of AIDS; and many specifics of current history. However, it is beyond the scope of this brief review to explore the validity of some of the more fanciful claims regarding these codes and to adequately review the serious caveats regarding their implications. *Sources: Bibliography Missler, Chuck, Cosmic Codes - Hidden Messages From the Edge of Eternity, Koinonia House, 1999*

Chapter 2

Prophecies of Jesus

This chapter provides a comprehensive examination of the prophecies concerning Jesus Christ in the Old Testament and their fulfillment in the New Testament. It begins with an introduction to Biblical prophecy, explaining its significance in establishing the credibility of scripture. The chapter then delves into the over 300 Messianic prophecies, focusing on key prophetic books like Isaiah, Psalms, Micah, and Zechariah. It systematically explores prophecies related to Jesus' birth, life, ministry, betrayal, death, resurrection, and ascension, providing specific scriptural references for each. The chapter includes a statistical analysis of the probability of these prophecies being fulfilled by chance.

Proof God Exists:

Indisputable Proof the Bible is True

God has left us with some indisputable evidence that He exists. The Old Testament prophets, who wrote certain books of the Bible, prophesied about specific events that would take place hundreds of years in the future. How is it possible for anyone to predict the future when it has not yet occurred? The prophets claimed that they received their information directly from God, who told them what would happen. Each prediction of the future is called a prophecy, and the Bible records hundreds of prophecies.

The prophets wrote down numerous predictions in the Scriptures, which came to pass hundreds of years later exactly as they prophesied. Bible prophecies that have already been fulfilled provide proof that God knows the future, can control earthly events, and that the Bible is true. Jesus prophesied to His disciples about His own crucifixion and resurrection so they would believe after it came to pass. In John 13:19; 14:29 He said "Now I tell you before it come, that, when it is come to pass, ye may believe that I am he."

In other words, God made prophecies, which are recorded in the Bible, and brought them to pass to give us evidence so we would believe in Him.

The prophets would die long before most of their predictions would come to pass, and time has proven them to be correct. Today, thanks to the meticulous work of scribes who carefully copied and preserved their predictions, we can look back on history and see that their prophecies were fulfilled exactly as they had proclaimed. This gives us proof that the Bible is inspired by God and we can therefore trust what it says.

Some Biblical Prophecies of the Future

The prophet Daniel gave some amazing details about how the future would unfold. In Daniel's prophecy of the Seventy Weeks (Daniel 9:24–27), he prophesied both the first and second comings of Christ. He also predicted that four world empires would successively arise over the span of hundreds of years— Babylon, then Medo-Persia, followed by Greece, and finally Rome (Daniel 2:31–45; 7:1–28). In Daniel

11:29–31, he predicted a king would arise and desecrate the temple in Jerusalem. This was fulfilled when Antiochus

Epiphanes slaughtered a pig on the altar of the Jewish temple and erected an altar to the god Zeus in 168 BC.

The prophet Micah lived over 700 years before Christ. God communicated a specific future event to this prophet, which was recorded in the Hebrew Bible: "But as for you, Bethlehem Ephrathah, too little to be among the clans of Judah, from you One will go forth for Me to be ruler in Israel. His goings forth are from long ago, from the days of eternity" (Micah 5:2). Out of all the cities in the world, Micah predicted that Bethlehem would be the location where the Messiah would be born. Jesus was born in Bethlehem, fulfilling the prophecy.

The prophet **Zechariah** predicted the Messiah would be betrayed for thirty pieces of silver, which was fulfilled when Judas betrayed Jesus. The prophet accurately described the exact number of pieces (30, not 29) and that the money would be of silver, not gold. He foretold that the money would be thrown in the sanctuary (not somewhere else) and used to buy a potter's field (Zech. 11:12–13; Matt. 27:3–7). He also prophesied that He would be pierced, which was fulfilled when the soldiers pierced Jesus' side with a spear (Zech. 12:10; John 19:34).

The prophet Isaiah, writing 700 years before Christ, predicted the Messiah would be scourged (Isaiah 53:5; John 19:1) and buried in a rich man's tomb (Isaiah 53:9; Matt. 27:60).

In Psalm 22, 1,000 years before Christ was born, King David described the crucifixion of Jesus, saying that His hands and feet would be pierced (Psalm 22:16; Luke 24:39), including the fact that they would cast lots for his garments, which was fulfilled when the soldiers cast lots for His outer garments and tunic (Psalm 22:18; John 19:23–24). David also predicted that the Messiah would be betrayed by a friend (Psalm 41:9; John 13:18), not a bone would be broken (Psalm 34:20; John 19:36), and He would rise from the dead (Psalm 16:10; Acts 2:31–32).

We have not begun to scratch the surface of all the biblical prophecies and their fulfillments. The fulfillment of each prophecy tells us that God oversaw the people and circumstances so that everything came together at just the right time. The fact that hundreds of Bible prophecies have been fulfilled exactly as foretold is irrefutable evidence that God not only exists, but that He is controlling history!

The prophecies below were made hundreds of years before Jesus was born and were fulfilled as foretold.

Prophecies and Fulfillments

Prophecies in the Old Testament about the Messiah	Approximate date prophesied	Fulfillment in the New Testament
He would be born of a virgin (Isaiah 7:14)	740-680 BC	Matt.1:18–25
He would be the Son of God (Psalm 2:7)	1,000 BC	Matt. 3:17
He would be a descendant of Abraham (Genesis 12:3, 22:18)	1,400 BC	Matt. 1:1

The Messiah would be a descendant of Isaac (Genesis 21:12)	1,400 BC	Matt.1:2 Luke 3:34

The Messiah would be a descendant of Jacob (Numbers 24:17)	1,400 BC	Matthew 1:2; Luke 3:34]
He would be from tribe of Judah (Genesis 49:10)	1,400 BC	Matthew 1:2; Luke 3:33

He would be from the family of Jesse (Isaiah 11:1)	740–680 BC	Matthew 1:6 Luke 3:32

Prophecies in the Old Testament about the Messiah	Approximate date prophesied	Fulfillment in the New Testament
The Messiah would be from house of David (Jeremiah 23:5)	627–580 BC	Matthew 1:1 Luke 3:31
He would be raised up as a prophet like Moses (Deuteronomy 18:15, 18)	1,400 BC	Acts 3:22 7:37
The Messiah would be born in Bethlehem (Micah 5:2)	722 BC	Matthew 2:1

After He was born, babies would be killed in Bethlehem (Jeremiah 31:15)	627–580 BC	Matthew 2:16–18
He would be called Immanuel, meaning "God with us" (Isaiah 7:14)	740–680 BC	Matthew 1:23
The Messiah would come from Galilee (Isaiah 9:1–2)	740–680 BC	Matthew 4:13–16
The Spirit of the Lord would be upon Him (Isa. 61:1)	740–680 BC	Luke 4:16–21; Matt. 12:17–18

He would be preceded by a messenger (Malachi 3:1)	430 BC	Matt. 11:10
He would do miracles (Isaiah 35:5–6)	740–680 BC	Matt. 11:2–5
Israel's king would ride into Jerusalem on donkey (Zechariah 9:9)	470 BC	Matt. 21:5–9 John 12:14
The Messiah would be welcomed with "Blessed is He who comes in the name of the Lord" (Psalm 118:26)	1000 BC	John 12:13

He would be hated for no reason (Psalm 35:19; 69:4)	1,000 BC	John 15:25
He would be rejected by the religious rulers (Psalm 118:22)	1000 BC	Matthew 21:42
He would be rejected by His own brothers (Psalm 69:8)	1,000 BC	John 7:5
He would be betrayed by a friend (Psalm 41:9)	1,000 BC	Matt. 10:4

Prophecies in the Old Testament about the Messiah	Approximate date prophesied	Fulfillment in the New Testament
His betrayer would eat bread with Him (Psalm 41:9)	1,000 BC	John 13:18, 26; Mark 14:18
He would be betrayed for money—30 pieces of silver (Zechariah 11:12)	470 BC	Matthew 26:15
Predicted exactly 30 pieces (not 29 or 31). The coins would be silver, not gold.	470 BC	Matthew 26:15

The money would be returned (Zechariah 11:12–13)	470 BC	Matthew 27:3
The money would be thrown in the house of the Lord (Zechariah 11:13)	470 BC	Matthew 27:5
The betrayal money would pay for a Potter's field (Zechariah 11:13)	470 BC	Matthew 27:7
He would be forsaken by the disciples (Zechariah 13:7)	470 BC	Matthew 26:31, 56
He would be silent before His accusers (Isaiah 53:7)	740–680 BC	Matthew 26:62–63

The Messiah would be mocked (Isa. 53:3)	740–680 BC	Matt. 27:29
]He would be beaten with a rod (Mic. 5:1)	722 BC	Mark 15:19
He would be spat upon in the face (Isa. 50:6)	740–680 BC	Mark 14:65
The Messiah would be wounded, bruised (Isaiah 53:5)	740–680 BC	Matt. 27:30; Luke 22:63
The Messiah would be scourged on His back (Isaiah 50:6, 53:5)	740–680 BC	John 19:1

His hands and feet would be pierced (Psalm 22:16)	1,000 BC	John 20:25
His garments would be divided (Psalm 22:18)	1,000 BC	John 19:23
They would cast lots for His clothing (Psalm 22:18)	1,000 BC	John 19:24
The Messiah would die with criminals (Isaiah 53:12)	740–680 BC	Mark 15:28; Luke 22:37

Prophecies in the Old Testament about the Messiah	Approximate date prophesied	Fulfillment in the New Testament
Those watching the crucifixion would wag their heads (Psalm 22:7; 109:25)	1,000 BC	Mark 15:29; Matt. 27:39
Those watching the crucifixion would mock Him for not saving Himself (Ps. 22:8)	1,000 BC	Matthew 27:41–43
He would pray for those crucifying Him (Isaiah 53:12)	740–680 BC	Luke 23:34

He would be given vinegar to drink (Ps. 69:21)	1,000 BC	Matt. 27:34
"Why hast thou forsaken me?" (Ps. 22:1)	1,000 BC	Matt. 27:46
"Into thine hand I commit my spirit" (Ps. 31:5)	1,000 BC	Luke 23:46
His side would be pierced (Zech. 12:10)	470 BC	John 19:34, 37
None of the Messiah's bones would be broken (Psalm 34:20)	1,000 BC	John 19:32–36

He would be buried in a rich man's tomb (Isaiah 53:9)	740–680 BC	Matthew 27:57–60
He would be dead for three days and three nights (Jonah 1:17)	760 BC	Matthew 12:40
He would descend into hell (Psalm 16:10; 49:15)	1,000 BC	Acts 2:27, 31; Eph. 4:9
The Messiah would be resurrected from dead (Psalm 16:10; 30:3)	1,000 BC	Acts 2:31; 13:33–35

Through His resurrection He would swallow up death in victory (Isaiah 25:8)	740–680 BC	1 Cor. 15:54
He would ascend into heaven (Psalm 68:18)	1,000 BC	Acts 1:9; Eph. 4:8–10
He would be seated at the right hand of the Father in heaven (Psalm 110:1)	1,000 BC	Acts 2:34–35; Col. 3:1
He would be a priest according to the order of Melchizedek (Psalm 110:4)	1,000 BC	Hebrews 5:6, 10; 6:20

The Messiah would be a light to the entire world, including non-Jews (Isaiah 42:6; 49:6)	740–680 BC	Luke 2:32; Acts 13:47; 26:23

Jesus fulfilled more than 300 prophecies in the Bible. Peter Stoner in *Science Speaks* (Chicago: Moody Press, 1963) calculated the probability of one man fulfilling 48 prophecies to be one in 10 to the 157th power.

> 1 in 10,000, 000, 000, 000, 000, 000, 000, 000, 000, 000, 000, 000, 000, 000, 000, 000, 000, 000, 000,
>
> 000, 000, 000, 000, 000, 000, 000, 000, 000, 000, 000, 000, 000, 000, 000, 000, 000, 000, 000, 000, 000, 000, 000, 000, 000, 000, 000, 000, 000, 000, 000, 000, 000, 000

This could not have happened by mere chance. Only a supernatural God could have inspired and fulfilled the prophecies in the Bible.

The evidence is so compelling that it should convince you beyond a reasonable doubt.
Source: Peter Stoner, Science Speaks, Making Life Count Ministries.

92 Prophecies of the Psalms Fulfilled in Jesus Christ

Psalm Prophecies	OT Scripture	NT Fulfilment
Christ would also be rejected by Gentiles.	Psalm 2:1	Acts 4:25-28
Political/religious leaders would conspire against Christ.	Psalm 2:2	Matthew 26:34 Mark 3:6

Christ would be King of the Jews.	Psalm 2:6	John 12:12-13 John 18:32
Christ would reveal that He was the Son of God.	Psalm 2:7	John 9:35-37
Christ would be raised from the dead and be crowned King.	Psalm 2:7	Acts 13:30-33 Romans 1:3-4
Christ would ask God for His inheritance.	Psalm 2:8	John 17:4-24

Christ would have complete authority over all things.	Psalm 2:8	Matthew 28:18 Hebrews 1:1-2

Christ would not acknowledge those who did not believe in Him.	Psalm 2:12	John 3:36
Infants would give praise to Christ.	Psalm 8:2	Matthew 21:15-16
Christ would have complete authority over all things.	Psalm 8:6	Matthew 28:18

Christ would be resurrected.	Psalm 16:81	Matthew 28:6 Acts 2:25-32
Christ's body would not see corruption (natural decay).	Psalm 16:81	Acts 13:35-37
Christ would be glorified into the presence of God.	Psalm 16:11	Acts 2:25-33
Christ would come for all people.	Psalm 18:49	Ephesians 3:4-6
Christ would cry out to God.	Psalm 22:1	Matthew 27:46

Christ would be forsaken by God at His crucifixion.	Psalm 22:1	Mark 15:34
Christ would pray without ceasing before His death.	Psalm 22:2	Matthew 26:38-39
Christ would be despised and rejected by His own.	Psalm 22:6	Luke 23:21-23
Christ would be made a mockery.	Psalm 22:7	Matthew 27:39
Unbelievers would say to Christ, "He trusted in God, let Him now deliver Him."	Psalm 22:8	Matthew 27:41-43

]Christ would know His Father from childhood.	Psalm 22:9	Luke 2:40
Christ would be called by God while in the womb.	Psalm 22:10	Luke 1:30-33
Christ would be abandoned by His disciples.	Psalm 22:11	Mark 14:50

Christ would be encompassed by evil spirits.	Psalm 22:12-13	Colossians 2:15

Christ's body would emit blood & water.	Psalm 22:14	John 19:34
Christ would be crucified.	Psalm 22:14	Matthew 27:35
Christ would thirst while dying.	Psalm 22:15	John 19:28
Christ would thirst just prior to His death.	Psalm 22:15	John 19:30
Christ would be observed by Gentiles at His crucifixion.	Psalm 22:16	Luke 23:36

Christ would be observed by Jews at His crucifixion.	Psalm 22:16	Matthew 27:41-43
Both Christ's hands and feet would be pierced.	Psalm 22:16	Matthew 27:38
Christ's bones would not be broken.	Psalm 22:17	John 19:32-33
Christ would be viewed by many during His crucifixion.	Psalm 22:17	Luke 23:35
Christ's garments would be parted among the soldiers.	Psalm 22:18	John 19:23-24

The soldiers would cast lots for Christ's clothes.	Psalm 22:18	John 19:23-24
Christ's atonement would enable believers to receive salvation.	Psalm 22:22	Hebrews 2:10-12 Matthew 12:50 John 20:14
Christ's enemies would stumble and fall.	Psalm 27:2	John 18:3-6
Christ would be accused by false witnesses.	Psalm 27:12	Matthew 26:59-61

Christ would cry out to God "Into thy hands I commend my spirit."	Psalm 31:5	Luke 23:46
Christ would come from the lineage of David.	Psalm 89:29	Matthew 1:1
Christ would come from the lineage of David.	Psalm 89:35-36	Matthew 1:1
Christ would be eternal.	Psalm 102:25-27	Revelation 1:8 Hebrews 1:10-12

Christ would be the creator of all things.	Psalm 102:25-27	John 1:3 Ephesians 3:9
Christ would calm the stormy sea.	Psalm 107:28-29	Matthew 8:24-26
Christ would be accused by many false witnesses.	Psalm 109:2	John 18:29-30
Christ would offer up prayer for His enemies.	Psalm 109:4	Luke 23:34
Christ's betrayer (Judas) would have a short life.	Psalm 109:8	Acts 1:16-18 John 17:12

Christ's betrayer would be replaced by another.	Psalm 109:8	Acts 1:20-26
Christ would be mocked by many.	Psalm 109:25	Mark 15:29-30
Christ would be Lord and King.	Psalm 110:1	Mat. 22:41-45

Christ would be exalted to the right hand of God.	Psalm 110:1	Mark 16:19 Mat. 22:41-46

Christ would be a Priest after the order of Melchisedek.	Psalm 110:4	Hebrews 6:17-20
Christ would be exalted to the right hand of God.	Psalm 110:5	1 Peter 3:21-22
Christ would be the "Stone" rejected by the builders (Jews).	Psalm 118:22	Mat.21:42-43
Christ would come in the name of the Lord.	Psalm 118:26	Matthew 21:9
Christ would come from the lineage of David.	Psalm 132:11	Matthew 1:1

Christ would come from the lineage of David.	Psalm 132:17	Matthew 1:1 Luke 1:68-70
There would be many attempts to kill Christ.	Psalm 31:13	Matthew 27:1
Christ would have no bones broken.	Psalm 34:20	John 19:32-33
Christ would be accused by many false witnesses.	Psalm 35:11	Mark 14:55-59
Christ would be hated without cause.	Psalm 35:19	John 18:19-23 John 15:24-25

Christ would be silent as a lamb before His accusers.	Psalm 38:13-14	Matthew 26:62-66
Christ would be God's sacrificial lamb for redemption of all mankind.	Psalm 40:6-8	Hebrews 10:10-13

Christ would reveal that the Hebrew scriptures were written of Him.	Psalm 40:6-8	Luke 24:44 John 5:39-40
Christ would do God's (His Father) will.	Psalm 40:7-8	John 5:30
Christ would not conceal His mission from believing people.	Psalm 40:9-10	Luke 4:16-21

Christ would be betrayed by one of His own disciples.	Psalm 41:9	Mark 14:17-18
Christ would communicate a message of mercy.	Psalm 45:2	Luke 4:22
Christ's throne would be eternal.	Psalm 45:6-7	Luke 1:31-33 Hebrews 1:8-9
Christ would be God.	Psalm 45:6-7	Hebrews 1:8-9
Christ would act with righteousness.	Psalm 45:6-7	John 5:30

Christ would be betrayed by one of His own disciples.	Psalm 55:12-14	Luke 22:47-48
Christ would ascend back into heaven.	Psalm 68:18	Luke 24:51 Ephesians 4:8
Christ would give good gifts unto believing men.	Psalm 68:18	Matthew 10:1 Ephesians 4:7-11
Christ would be hated and rejected without cause.	Psalm 69:4	Luke 23:13-22

Christ would be condemned for God's sake.	Psalm 69:7	Mat. 26:65-66
Christ would be rejected by the Jews.	Psalm 69:8	john 1:11
Christ's very own brothers would reject Him.	Psalm 69:8	John 7:3-5
Christ would become angry due to unethical practices by the Jews in the temple.	Psalm 69:9	John 2:13-17

Christ would be condemned for God's sake.	Psalm 69:9	Romans 15:3
Christ's heart would be broken.	Psalm 69:20	John 19:34
Christ's disciples would abandon Him just prior to His death.	Psalm 69:20	Mark 14:33-41
Christ would be offered gall mingled with vinegar while dying.	Psalm 69:21	Matthew 27:34
Christ would thirst while dying.	Psalm 69:21	John 19:28

The potters field would be uninhabited (Field of Blood)	Psalm 69:25	Acts 1:16-20
Christ would teach in parables.	Psalm 78:2	Mat.13:34-35
Christ would be exalted to the right hand of God.	Psalm 80:17	Acts 5:31
Christ would come form the lineage of David.	Psalm 89:3-4	Matthew 1:1
Christ would call God His Father.	Psalm 89:26	Matthew 11:27
Christ would be God's only "begotten" Son.	Psalm 89:27	Mark 16:6

Fulfilled Prophecy: Evidence for the Reliability of the Bible

Unique among all books ever written, the Bible accurately foretells specific events-in detail-many years, sometimes centuries, before they occur. Approximately 2,500 prophecies appear in the pages of the Bible, about 2,000 of which already have been fulfilled to the letter—no errors.

(The remaining 500 or so reach into the future and may be seen unfolding as days go by.) Since the probability for any one of these prophecies having been fulfilled by chance averages less than one in ten (figured very conservatively) and since the prophecies are for the most part independent of one another, the odds for all these prophecies having been fulfilled by chance without error is less than one in 10^{2000} (that is 1 with 2,000 zeros written after it)!

God is not the only one, however, who uses forecasts of future events to get people's attention. Satan does, too. Through clairvoyants (such as Jeanne Dixon and Edgar Cayce), mediums, spiritists, and others, come remarkable predictions, though rarely with more than about 60 percent accuracy, never with total accuracy. Messages from Satan, furthermore, fail to match the detail of Bible prophecies, nor do they include a call to repentance.

The acid test for identifying a prophet of God is recorded by Moses in Deuteronomy 18:21-22. According to this Bible passage (and others), God's prophets, as distinct from Satan's spokesmen, are 100 percent accurate in their predictions. There is *no* room for error.

As economy does not permit an explanation of all the Biblical prophecies that have been fulfilled, what follows in a discussion of a few that exemplify the high degree of specificity, the range of projection, and/or the "supernature" of the predicted events. Readers are encouraged to select others, as well, and to carefully examine their historicity.

(1) Some time before 500 BC, the prophet Daniel proclaimed that Israel's long-awaited Messiah would begin his public ministry 483 years after the issuing of a decree to restore and rebuild Jerusalem (Daniel 9:25-26). He further predicted that the Messiah would be "cut off," killed, and that this event would take place prior to a second destruction of Jerusalem. Abundant documentation shows that these prophecies were perfectly fulfilled in the life (and crucifixion) of Jesus Christ. The decree regarding the restoration of Jerusalem was issued by Persia's King Artaxerxes to the Hebrew priest Ezra in 458 BC, 483 years later the ministry of Jesus Christ began in Galilee. (Remember that due to calendar changes, the date for the start of Christ's ministry is set by most historians at about AD 26. Also note that from 1 BC to AD 1 is just one year.) Jesus' crucifixion occurred only a few years later, and about four decades later, in AD 70 came the destruction of Jerusalem by Titus.

(Probability of chance fulfillment = 1 in 10^5.)*

(2) In approximately 700 BC, the prophet Micah named the tiny village of Bethlehem as the birthplace of Israel's Messiah (Micah 5:2). The fulfillment of this prophecy in the birth of Christ is one of the most widely known and widely celebrated facts in history.

(Probability of chance fulfillment = 1 in 10^5.)

(3) In the fifth century BC, a prophet named Zechariah declared that the Messiah would be betrayed for the price of a slave—thirty pieces of silver, according to Jewish law-and also that this money would be used to buy a burial ground for Jerusalem's poor foreigners (Zechariah 11:12-13). Bible writers and secular historians both record thirty pieces of silver as the sum paid to Judas Iscariot for betraying Jesus, and they indicate that the money went to purchase a "potter's field," used—just as predicted—for the burial of poor aliens (Matthew 27:3-10).

(Probability of chance fulfillment = 1 in 10^{11}.)

(4) Some 400 years before crucifixion was invented, both Israel's King David and the prophet Zechariah described the Messiah's death in words that perfectly depict that mode of execution. Further, they said that the body would be pierced and that none of the bones would be broken, contrary to customary procedure in cases of crucifixion (Psalm 22 and 34:20; Zechariah 12:10). Again, historians and New Testament writers confirm the fulfillment: Jesus of Nazareth died on a Roman cross, and his extraordinarily quick death eliminated the need for the usual breaking of bones. A spear was thrust into his side to verify that he was, indeed, dead.

(Probability of chance fulfillment = 1 in 10^{13}.)

---

(5) The prophet Isaiah foretold that a conqueror named Cyrus would destroy seemingly impregnable Babylon and subdue Egypt along with most of the rest of the known world. This same man, said Isaiah, would decide to let the Jewish exiles in his territory go free without any payment of ransom (Isaiah 44:28; 45:1; and 45:13). Isaiah made this prophecy 150 years before Cyrus was born, 180 years before Cyrus performed any of these feats (and he did, eventually, perform them all), and 80 years before the Jews were taken into exile.

(Probability of chance fulfillment = 1 in 10^{15}.)

(6) Mighty Babylon, 196 miles square, was enclosed not only by a moat, but also by a double wall 330 feet high, each part 90 feet thick. It was said by unanimous popular opinion to be indestructible, yet two Bible prophets declared its doom. These prophets further claimed that the ruins would be avoided by travelers, that the city would never again be inhabited, and that its stones would not even be moved for use as building material (Isaiah 13:17-22 and Jeremiah 51:26, 43). Their description is, in fact, the well-documented history of the famous citadel.

(Probability of chance fulfillment = 1 in 10^9.)

---

(7) The exact location and construction sequence of Jerusalem's nine suburbs was predicted by Jeremiah about 2600 years ago. He referred to the time of this building project as "the last days," that is, the time period of Israel's second rebirth as a nation in the land of Palestine (Jeremiah 31:38-40). This rebirth became history in 1948, and the construction of the nine suburbs has gone forward precisely in the locations and in the sequence predicted.

(Probability of chance fulfillment = 1 in 10^{18}.)

---

(8) The prophet Moses foretold (with some additions by Jeremiah and Jesus) that the ancient Jewish nation would be conquered twice and that the people would be carried off as slaves each time, first by the Babylonians (for a period of 70 years), and then by a fourth world kingdom (which we know as Rome). The second conqueror, Moses said, would take the Jews captive to Egypt in ships, selling them or giving them away as slaves to all parts of the world. Both of these predictions were fulfilled to the letter, the first in 607 BC and the second in AD 70. God's spokesmen said, further, that the Jews would remain scattered throughout the entire world for many generations, but without becoming assimilated by the peoples or of other nations, and that the Jews would one day return to the land of Palestine to re-establish for a second time their nation (Deuteronomy 29; Isaiah 11:11-13; Jeremiah 25:11; Hosea 3:4-5 and Luke 21:23-24).

This prophetic statement sweeps across 3,500 years of history to its complete fulfillment—in our lifetime.

(Probability of chance fulfillment = 1 in 10^{20}.)

---

(9) Jeremiah predicted that despite its fertility and despite the accessibility of its water supply, the land of Edom (today a part of Jordan) would become a barren, uninhabited wasteland (Jeremiah 49:15-20; Ezekiel 25:12-14). His description accurately tells the history of that now bleak region.

(Probability of chance fulfillment = 1 in 10^5.)

(10) Joshua prophesied that Jericho would be rebuilt by one man. He also said that the man's eldest son would die when the reconstruction began and that his youngest son would die when the work reached completion (Joshua 6:26). About five centuries later this prophecy found its fulfillment in the life and family of a man named Hiel (1 Kings 16:33-34).

(Probability of chance fulfillment = 1 in 10^7).

---

(11) The day of Elijah's supernatural departure from Earth was predicted unanimously—and accurately, according to the eye-witness account—by a group of fifty prophets (2 Kings 2:3-11).

(Probability of chance fulfillment = 1 in 10^9).

---

(12) Jahaziel prophesied that King Jehoshaphat and a tiny band of men would defeat an enormous, well-equipped, well-trained army without even having to fight. Just as predicted, the King and his troops stood looking on as their foes were supernaturally destroyed to the last man (2 Chronicles 20).

(Probability of chance fulfillment = 1 in 10^8).

---

(13) One prophet of God (unnamed, but probably Shemiah) said that a future king of Judah, named Josiah, would take the bones of all the occultic priests (priests of the "high places") of Israel's King Jeroboam and burn them on Jeroboam's altar (1 Kings 13:2 and 2 Kings 23:15-18). This event occurred approximately 300 years after it was foretold.

(Probability of chance fulfillment = 1 in 10^{13}).

Since these thirteen prophecies cover mostly separate and independent events, the probability of chance occurrence for all thirteen is about 1 in 10^{138} (138 equals the sum of all the exponents of 10 in the probability estimates above). For the sake of putting the figure into perspective, this probability can be compared to the statistical chance that the second law of thermodynamics will be reversed in a given situation (for example, that a gasoline engine will refrigerate itself during its combustion cycle or that heat will flow from a cold body to a hot body)—that chance = 1 in 10^{80}. Stating it simply, based on these thirteen prophecies alone, the Bible record may be said to be vastly more reliable than the second law of thermodynamics. Each reader should feel free to make his own reasonable estimates of probability for the chance fulfillment of the prophecies cited here. In any case, the probabilities deduced still will be absurdly remote.

Given that the Bible proves so reliable a document, there is every reason to expect that the remaining 500 prophecies, those slated for the "time of the end," also will be fulfilled to the last letter. Who can afford to ignore these coming events, much less miss out on the immeasurable blessings offered to anyone and everyone who submits to the control of the Bible's author, Jesus Christ? Would a reasonable person take lightly God's warning of judgment for those who reject what they know to be true about Jesus Christ and the Bible, or who reject Jesus' claim on their lives?

*The estimates of probability included herein come from a group of secular research scientists. As an example of their method of estimation, consider their calculations for this first prophecy cited:

- Since the Messiah's ministry could conceivably begin in any one of about 5000 years, there is, then, one chance in about 5,000 that his ministry could begin in AD 26.

- Since the Messiah is God in human form, the possibility of his being killed is considerably low, say less than one chance in 10.

- Relative to the second destruction of Jerusalem, this execution has roughly an even chance of occurring before or after that event, that is, one chance in 2.

Hence, the probability of chance fulfillment for this prophecy is 1 in 5,000 x 10 x 2, which is 1 in 100,000, or 1 in 10^5.

Source: Hugh Ross, Reasons to Believe.

Betrayal Of Jesus Foretold In Zechariah (KJV)

Zechariah 11:12-13 offers an intriguing account regarding 30 silver coins, bringing to mind the betrayal of Jesus by Judas Iscariot. The New Testament identifies this as a Messianic prophecy, which found its fulfillment in Jesus Christ.

The verses read, "I told them, 'If you think it best, give me my pay; but if not, keep it.' So they paid me thirty pieces of silver. And the Lord said to me, 'Throw it to the potter'—the handsome price at which they priced me! So I took the thirty pieces of silver and threw them into the house of the Lord to the potter."

Earlier, Zechariah had been commanded to watch a flock of sheep doomed to slaughter (Zechariah 11:4). He obeyed, using two shepherd's staffs that he named Favor and Union (verse 7). Within a month, Zechariah fired the three shepherds working under him (verse 8). Then Zechariah abandoned the flock and broke his staff named Favor. Observers realized these actions were "the word of the LORD" (verse 11). The Lord would remove His favor from His people, allowing them to be harried by their enemies (verse 6).

In verses 12-13 Zechariah tells his employers to pay him his wages if they saw fit to do so. They pay him 30 pieces of silver, the price of a slave (Exodus 21:32), as an insult to Zechariah. The prophet sarcastically calls it a "handsome price." God then commands Zechariah to give the coins to the potter in the house (or temple) of the Lord.

The corresponding passage in the New Testament is in Matthew 27. Judas is filled with remorse for betraying the Lord, and he tries to return the thirty pieces of silver to the chief priests (verse 3). When the elders refuse to accept the money, Judas throws the coins into the temple and leaves and hangs himself (verses 4-5). Not wanted to put "blood money" into the treasury, the priests use it to buy a potter's field (verses 6-7). "Then what was spoken by Jeremiah the prophet was fulfilled: 'They took the thirty silver coins, the price set on him by the people of Israel, and they used them to buy the potter's field, as the Lord commanded me'"

A Messianic Prophecy

Psalm 22 (KJV)

1 My God, my God, why hast thou forsaken me? why art thou so far from helping me, and from the words my roaring?

2 O my God, I cry in the day time, but thou hearest not; and in the night season, and am not silent.

3 But thou art holy, O thou that inhabitest the praises of Israel.

4 Our fathers trusted in thee: they trusted, and thou didst deliver them.

5 They cried unto thee, and were delivered: they trusted in thee, and were not confounded.

6 But I am a worm, and no man; a reproach of men, and despised of the people

7 All they that see me laugh me to scorn: they shoot out the lip, they shake the head, saying,

8 He trusted on the Lord that he would deliver him: let him deliver him, seeing he delighted in him.

9 But thou art he that took me out of the womb: thou didst make me hope when I was upon my mother's breasts.

10 I was cast upon thee from the womb: thou art my God from my mother's belly.

11 Be not far from me; for trouble is near; for there is none to help.

12 Many bulls have compassed me: strong bulls of Bashan have beset me round.

13 They gaped upon me with their mouths, as a ravening and a roaring lion.

14 I am poured out like water, and all my bones are out of joint: my heart is like wax; it is melted in the midst of my bowels.

15 My strength is dried up like a potsherd; and my tongue cleaveth to my jaws; and thou hast brought me into the dust of death.

16 For dogs have compassed me: the assembly of the wicked have inclosed me: they pierced my hands and my feet.

17 I may tell all my bones: they look and stare upon me.

18 They part my garments among them, and cast lots upon my vesture.

19 But be not thou far from me, O Lord: O my strength,

haste thee to help me.

20 Deliver my soul from the sword; my darling from the power of the dog.

21 Save me from the lion's mouth: for thou hast heard me from the horns of the unicorns.

22 I will declare thy name unto my brethren: in the midst of the congregation will I praise thee.

23 Ye that fear the Lord, praise him; all ye the seed of Jacob, glorify him; and fear him, all ye the seed of Israel.

24 For he hath not despised nor abhorred the affliction of the afflicted; neither hath he hid his face from him; but when he cried unto him, he heard.

25 My praise shall be of thee in the great congregation: I will pay my vows before them that fear him.

26 The meek shall eat and be satisfied: they shall praise the Lord that seek him: your heart shall live for ever.

27 All the ends of the world shall remember and turn unto the Lord: and all the kindreds of the nations shall worship before thee.

28 For the kingdom is the Lord's: and he is the governor among the nations.

29 All they that be fat upon earth shall eat and worship: all

they that go down to the dust shall bow before him: and none can keep alive his own soul.

30 A seed shall serve him; it shall be accounted to the Lord for a generation.

31 They shall come, and shall declare his righteousness unto a people that shall be born, that he hath done this.

A Messianic Prophecy

Isaiah 53 (KJV)

1 Who hath believed our report? and to whom is the arm of the Lord revealed?

2 For he shall grow up before him as a tender plant, and as a root out of a dry ground: he hath no form nor comeliness; and when we shall see him, there is no beauty that we should desire him.

3 He is despised and rejected of men; a man of sorrows, and acquainted with grief: and we hid as it were our faces from him; he was despised, and we esteemed him not.

4 Surely he hath borne our griefs, and carried our sorrows: yet we did esteem him stricken, smitten of God, and afflicted.

5 But he was wounded for our transgressions, he was bruised for our iniquities: the chastisement of our peace was upon him; and with his stripes we are healed.

6 All we like sheep have gone astray; we have turned every one to his own way; and the Lord hath laid on him the iniquity of us all.

7 He was oppressed, and he was afflicted, yet he opened not his mouth: he is brought as a lamb to the slaughter, and as a sheep before her shearers is dumb, so he openeth not his mouth.

8 He was taken from prison and from judgment: and who shall declare his generation? for he was cut off out of the land of the living: for the transgression of my people was he stricken.

9 And he made his grave with the wicked, and with the rich in his death; because he had done no violence, neither was any deceit in his mouth.

10 Yet it pleased the Lord to bruise him; he hath put him to grief: when thou shalt make his soul an offering for sin, he shall see his seed, he shall prolong his days, and the pleasure of the Lord shall prosper in his hand.

11 He shall see of the travail of his soul, and shall be satisfied: by his knowledge shall my righteous servant justify many; for he shall bear their iniquities.

12 Therefore will I divide him a portion with the great, and he shall divide the spoil with the strong; because he hath poured out his soul unto death: and he was numbered with the transgressors; and he bare the sin of many, and made intercession for the transgressors.

A Messianic Prophecy

Psalm 2 (KJV)

1 Why do the heathen rage, and the people imagine a vain thing?

2 The kings of the earth set themselves, and the rulers take counsel together, against the Lord, and against his anointed, saying,

3 Let us break their bands asunder, and cast away their cords from us.

4 He that sitteth in the heavens shall laugh: the Lord shall have them in derision.

5 Then shall he speak unto them in his wrath, and vex them in his sore displeasure.

6 Yet have I set my king upon my holy hill of Zion.

7 I will declare the decree: the Lord hath said unto me, Thou art my Son; this day have I begotten thee.

8 Ask of me, and I shall give thee the heathen for thine inheritance, and the uttermost parts of the earth for thy possession.

9 Thou shalt break them with a rod of iron; thou shalt dash them in pieces like a potter's vessel.

10 Be wise now therefore, O ye kings: be instructed, ye judges of the earth.

11 Serve the Lord with fear, and rejoice with trembling.

12 Kiss the Son, lest he be angry, and ye perish from the way, when his wrath is kindled but a little. Blessed are all they that put their trust in him.

I AM The Bread of Life (KJV)

God told the Israelites to eat unleavened bread during Passover as a commemoration of their exodus from Egyptian bondage. In the Bible, leaven is almost always symbolic of sin.

Jesus was born and laid in a feeding trough where animals eat grain (grain has to die to give life to a crop) he then referred to himself as the "Bread of Life" and died to give life to all of us.

1 Corinthians 15:36 Thou fool, that which thou sowest is not quickened, except it die

John 6:35 And Jesus said unto them I am the bread of life: he that cometh to me shall never hunger; and he that believeth on me shall never thirst.

John 12:24-26 Verily, verily, I say unto you, Except a corn of wheat fall into the ground and die, it abideth alone: but if it die, it bringeth forth much fruit. 25 He that loveth his life shall lose it; and he that hateth his life in this world shall keep it unto life eternal. 26 If any man serve me, let him follow me; and where I am, there shall also my servant be: if any man serve me, him will my Father honour.

A grain of wheat, though containing in itself the germs of life, would remain alone, and not really live unless it fell to the earth. Then the life-germs would burst forth, and the single grain, in its own death, would give life to blade, and stalk, and ear of corn. Its death then was the true life, for it released the inner life-power which the husk before held captive; and this life-power multiplying itself in successive grains would clothe the whole field with a harvest of much fruit.

Revelation 19:8 And to her was granted that she should be arrayed in fine linen, clean and white: for the fine linen is the righteousness of saints.

Jesus was laid in a manger which facilitated a metaphor=Newborn Jesus, the Bread of Life, lay in a feeding trough In the little town of Bethlehem, which means "the house of bread" In Hebrew. No other food has the power to give us everlasting life (John 6:51). No other nourishment can yield lasting joy (Jeremiah 15:16). Those who feast on this bread will never be hungry again (John 6:35).

But thou, Bethlehem Ephratah, though thou be little among the thousands of Judah, yet out of thee shall he come forth unto me that is to be ruler in Israel; whose goings forth have been from of old, from everlasting. (Micah 5:2)

It was foretold that Jesus would be born in Bethlehem, meaning House of Bread. This means that the name of His birthplace had always foretold the coming of the Bread of Life (John 6:35)

Bethlehem was also named Ephrath or Ephrathah, both meaning fruitful. This is not a coincidence either. To properly understand why say this, we must go back to a time about 1500 years before Jesus was born. In those days, God had promised Jacob that he would become fruitful (Genesis 35:11) and he had 13 children; the first generation of the people of Israel.

The book of Genesis explicitly points us to the fact that Jacob's beloved wife Rachel had died on their way to Bethlehem while giving birth to their youngest son due to hard labor (Genesis 35:16-20). The fact that Jacob's lastborn was born right before reaching Bethlehem is meaningful. Because at the time of Jesus's birth, being a descendant of Jacob was one of the ways in which the Israelites put confidence in the flesh (Philippians 3:3-5). They focused on physical fruit, rather than spiritual fruit. They were impressed with physical bread from heaven, supposedly coming from Moses, rather than the true Bread, The Bread of Life given by God (John 6:28-35).

Why was Jesus wrapped in cloths? And this shall be a sign unto you; Ye shall find the babe wrapped in swaddling clothes, lying in a manger. (Luke 2:12)

Jesus was swaddled with strips of cloth, presumably linen. In

Exodus and Leviticus, we can read many examples of holy things being wrapped or covered in linen. The curtains in the temple were made from linen (which has prophetic meaning itself) and the anointed priests were fully clothed in linen. These linen clothes, especially the underwear, had to always be worn by the priest and his offspring when ministering near God (Exodus 28:42-43, Ezekiel 44:17-19). The Levitical priests couldn't come to God without their linen clothes and a sacrificial animal (Leviticus 16:3-4), but Jesus was the sacrifice Himself.

As you know, Jesus came to fulfill the Law and bring a better covenant. He was born to be our great High Priest forever, offering an eternal sacrifice for our sins before God (Hebrews 4:14, 7:20-28). He ended His fulfillment of the Law by leaving the linen cloths behind, in the grave, where the dead belong. He is alive, and we are made alive through Him, no longer needing these superficial "dead" elements to become holy. All we need is His living Spirit, who guides us into holiness.

Why gold, myrrh, and frankincense?

Matthew 2:11 And when they were come into the house, they saw the young child with Mary his mother, and fell down, and worshipped him: and when they had opened their treasures, they presented unto him gifts; gold, and frankincense and myrrh.."

If you understand the previous connection between linen and priestly holiness, then you can also see why the wise men brought gold, and frankincense, and myrrh: frankincense, and myrrh: These three gifts weren't only seen as precious in those days, they were also brought to the priest as tithes and offerings.

Gold was used abundantly in and on the temple, and the temple is symbolic of Jesus (which is a whole other topic).

Frankincense was used on grain offerings and the Bread of the Presence (Leviticus 2, 24:7). Jesus became our grain offering and our Bread.

The scent of myrrh was greatly appreciated, and it was used as an ingredient to make sacred anointing oil, which was to be used only to anoint priests (Exodus 30:23-32). Jesus would become our High Priest. Myrrh was also traditionally used as an ingredient for embalming, and so it was used on Jesus's body when it was wrapped in linen cloths and buried (John 19:39-40).

 Adam and Eve were given the plants and fruits to eat in the Garden of Eden It was only after they sinned that reference is made to tilling the soil and growing grains, and this reference was mingled with a reference to death when God told Adam: "By the sweat of your brow you will eat your food until you return to the ground, since from it you were taken; for dust you are and to dust you will return."

Chapter 3

Jesus is God

The chapter presents Biblical evidence for the deity of Jesus Christ through a systematic examination of scripture references. It begins with John 1:1, establishing Jesus as the Word who was both with God and was God.

The chapter progresses through key verses including Thomas's declaration 'My Lord and my God' (John 20:28), Jesus's statement 'I and my Father are one' (John 10:30), and His claim 'Before Abraham was, I am' (John 8:58).

Additional evidence is presented from Paul's writings, including Colossians 2:9 ('in Him dwells all the fullness of the Godhead bodily') and Philippians 2:6 (Jesus's equality with God). The chapter concludes with Old Testament prophecies like Isaiah 9:6 that point to Jesus's divine nature, creating a comprehensive biblical case for Jesus's deity.

The Deity of Jesus in the Bible (KJV)

John 1:1 In the beginning was the Word and the Word was with God and the Word was God.

John 20:28 And Thomas answered and said unto him, my Lord, and my God.

Romans 9:5 Whose are the fathers and of whom as concerning the flesh Christ came, who is over all, God blessed for ever. Amen.

Philippians 2:6 Who, being in the form of God, thought it not robbery to be equal with God.

Titus 2:13 Looking for that blessed hope, and the glorious appearing of the great God and our savior Jesus Christ.

1 John 5:20 And we know that the Son of God is come, and hath given us an understanding, that we may know him that is true, and we are in him that is true, even in his son Jesus Christ. This is the true God, and eternal life.

1 Corinthians 8:6 Yet for us there is one God, The Father, from whom are all things for whom we exist, and one Lord, Jesus Christ, through whom are all things, and though whom we exist.

Romans 9:5 Whose are the fathers, and of whom as concerning the flesh Christ came, who is over all, God blessed for ever. Amen.

John 1:14 And the Word was made flesh, and dwelt among us, (and we beheld his glory, the glory as of the only begotten of the Father,) full of grace and truth.

John 10:30 I and my Father are one.

Mark 12:29 And Jesus answered him, The first of all the commandments is, Hear, O Israel; The Lord our God is one Lord.

John 8:58 Jesus said unto them, Verily, verily, I say unto you, Before Abraham was, I am.

Isaiah 9:60 For unto us a child is born, unto us a son is given: and the government shall be upon his shoulder: and his name shall be called Wonderful, Counsellor, The mighty God, The everlasting Father, The Prince of Peace.

Colossians 2:90 For in him dwelleth all the fulness of the Godhead bodily.

Hebrews 1:8 But unto the Son he saith, Thy throne, O God, is for ever and ever: a sceptre of righteousness is the sceptre of thy kingdom.

Matthew 1:23 Behold, a virgin shall be with child, and shall bring forth a son, and they shall call his name Emmanuel, which being interpreted is, God with us.

But to us there is but one God, the Father, of whom are all things, and we in him; and one Lord Jesus Christ, by whom are all things, and we by him.

John 5:18 Therefore the Jews sought the more to kill him, because he not only had broken the sabbath, but said also that God was his Father, making himself equal with God.

Matthew 4:1-7- 1 Then was Jesus led up of the Spirit into the wilderness to be tempted of the devil.

2 And when he had fasted forty days and forty nights, he was afterward an hungred.

3 And when the tempter came to him, he said, If thou be the Son of God, command that these stones be made bread.

4 But he answered and said, It is written, Man shall not live by bread alone, but by every word that proceedeth out of the mouth of God.

5 Then the devil taketh him up into the holy city, and setteth him on a pinnacle of the temple,

6 And saith unto him, If thou be the Son of God, cast thyself down: for it is written, He shall give his angels charge concerning thee: and in their hands they shall bear thee up, lest at any time thou dash thy foot against a stone.

7 Jesus said unto him, It is written again, Thou shalt not

tempt the Lord thy God.

8 Again, the devil taketh him up into an exceeding high mountain, and sheweth him all the kingdoms of the world, and the glory of them;

9 And saith unto him, All these things will I give thee, if thou wilt fall down and worship me.

10 Then saith Jesus unto him, Get thee hence, Satan: for it is written, Thou shalt worship the Lord thy God, and him only shalt thou serve.

John 1:1-3 In the beginning was the Word, and the Word was with God, and the Word was God.

2 The same was in the beginning with God.

3 All things were made by him; and without him was not any thing made that was made.

Isaiah 9:6 For unto us a Child is born, Unto us a Son is given; And the government will be upon His shoulder. And His name will be called Wonderful, Counselor, Mighty God, Everlasting Father, Prince of Peace."

Isaiah 43:10,11 You are My witnesses, says the Lord, And My servant whom I have chosen, That you may know and believe Me, and understand that I am He. Before Me there was no God formed, Nor shall there be after Me. I, even I, am the Lord, and besides Me there is no Savior.

Revelation 1:17 And when I saw him, I fell at his feet as dead. And he laid his right hand upon me, saying unto me, Fear not; I am the first and the last:

Isaiah 44:6 Thus saith the Lord the King of Israel, and his redeemer the Lord of hosts; I am the first, and I am the last; and beside me there is no God.

2 Peter 1:1 To those who have obtained like precious faith with us by the righteousness of our God and Savior Jesus Christ

Isaiah 44:24 (God created the world by Himself alone)

John 1:3 Colossians 1:16 - (Jesus made all things)

John 5:17,18 My Father has been working until now, and I have been working." Therefore the Jews sought to kill Him, because He not only broke the Sabbath, but also said that God was His Father, making Himself equal with God.

John 5:23 That all should honor the Son just as they honor the Father. He who does not honor the Son does not honor the Father who sent Him.

John 8:24 Therefore I said to you that you will die in your sins; for if you do not believe that I AM [He], you will die in your sins.

John 8:58 Then Jesus said to them, Most assuredly, I say to you, before Abraham was, I AM."

John 10:30-33 Jesus answered them, I and My Father are one. Then the Jews took up stones again to stone Him. Jesus answered them, Many good works I have shown you from My Father. For which of those works do you stone Me?

The Jews answered Him, saying, For a good work we do not stone You, but for blasphemy, and because You, being a Man, make Yourself God.

John 14:6-7 Jesus said to him, I AM the way, the truth, and the Life. No one comes to the Father except through Me.

John 14:9-11 Jesus said to him, Have I been with you so long and yet you have not known Me, Philip? He who has seen Me has seen the Father; so how can you say, 'Show us the Father'?"

John 20:28 And Thomas answered and said to Him, My Lord and my God!

4 Acts 4:12 Nor is there salvation in any other, for there is no other name under heaven given among men by which we must be saved.

Acts 20:28 Take heed therefore unto yourselves, and to all the flock, over the which the Holy Ghost hath made you overseers, to feed the church of God, which he hath purchased with his own blood.

Philippians 2:5-7 Let this mind be in you which was also in Christ Jesus, who, being in the form of God, did not consider it robbery to be equal with God, but made Himself of no reputation, taking the form of a bond-servant, and coming in the likeness of men."

Colossians 2:9 For in Him dwells all the fullness of the Godhead bodily."

1 Timothy 3:16 And without controversy great is the mystery of godliness: God was manifested in e flesh, Justified in the Spirit, Seen by angels, Preached among the Gentiles, Believed on in the world, Received up in glory."

Titus 2:13 looking for the blessed hope and glorious appearing of our great God and Savior Jesus Christ

Hebrews 1:8-9 But to the Son He says: Your throne, O God, is forever and ever; A scepter of righteousness is the scepter of Your kingdom. You have loved righteousness and hated lawlessness; Therefore God, Your God, has anointed You with the oil of gladness more than Your companions.

2 John 1:7 For many deceivers have gone out into the world who do not confess Jesus Christ as coming in the flesh. This is a deceiver and an antichrist.

Revelation 1:8 I am the Alpha and the Omega, the Beginning and the End, says the Lord, "who is and who was and who is to come, the Almighty.

Revelation 22:13 I AM the Alpha and the Omega, the Beginning and the End, the First and the Last.

Revelation 22:16 I, Jesus, have sent My angel to testify to you these things in the churches.

1 Timothy 6:14-16 Our Lord Jesus Christ's appearing, which He will manifest in His own time, He who is the blessed and only Potentate, the King of kings and Lord of lords, who alone has immortality, dwelling in unapproachable light, whom no man has seen or can see, to whom be honor and everlasting power. Amen.

Hebrews 2:17-18 Therefore, in all things He had to be made like His brethren, that He might be a merciful and faithful High Priest in things pertaining to God, to make propitiation for the sins of the people. For in that He Himself has suffered, being tempted, He is able to aid those who are tempted.

John 15:13 Greater love has no one than this, than to lay down one's life for his friends.

Romans 5:8 But God demonstrates His own love toward us, in that while we were still sinners, Christ died for us.

Chapter 4

The Historicity of Jesus

Historical Documents That Prove Jesus Lived and Was Crucified

This chapter examines extra-biblical historical evidence for Jesus's existence and crucifixion. It begins with Josephus's account in 'The Antiquities of the Jews' (Book XVIII, Chapter 3), which describes Jesus as a wise man who performed wonderful works and was crucified under Pontius Pilate.

The chapter then analyzes Tacitus's reference to 'Christus' in his Annals (written c. AD 116), providing Roman historical confirmation of Jesus's execution.

The account from Thallus (AD 52) regarding the darkness at Jesus's crucifixion is also examined. Each historical source is analyzed for authenticity and historical context, demonstrating that Jesus's existence and crucifixion are well-documented historical facts, not merely religious claims.

The Antiquities of the Jews

by Josephus, translated by William Whiston

Book XVIII

Chapter 3

SEDITION OF THE JEWS AGAINST PONTIUS PILATE. CONCERNING CHRIST, AND WHAT BEFELL PAULINA AND THE JEWS AT ROME,

1. BUT now Pilate, the procurator of Judea, removed the army from Cesarea to Jerusalem, to take their winter quarters there, in order to abolish the Jewish laws. So he introduced Caesar's effigies, which were upon the ensigns, and brought them into the city; whereas our law forbids us the very making of images; on which account the former procurators were wont to make their entry into the city with such ensigns as had not those ornaments. Pilate was the first who brought those images to Jerusalem, and set them up there; which was done without the knowledge of the people, because it was done in the night time; but as soon as

they knew it, they came in multitudes to Cesarea, and interceded with Pilate many days that he would remove the images; and when he would not grant their requests, because it would tend to the injury of Caesar, while yet they persevered in their request, on the sixth day he ordered his soldiers to have their weapons privately, while he came and sat upon his judgment-seat, which seat was so prepared in the open place of the city, that it concealed the army that lay ready to oppress them; and when the Jews petitioned him again, he gave a signal to the soldiers to encompass them routed, and threatened that their punishment should be no less than immediate death, unless they would leave off disturbing him, and go their ways home. But they threw themselves upon the ground, and laid their necks bare, and said they would take their death very willingly, rather than the wisdom of their laws should be transgressed; upon which Pilate was deeply affected with their firm resolution to keep their laws inviolable, and presently commanded the images to be carried back from Jerusalem to Cesarea.

2. But Pilate undertook to bring a current of water to Jerusalem, and did it with the sacred money, and derived the origin of the stream from the distance of two hundred furlongs. However, the Jews [8] were not pleased with what had been done about this water; and many ten thousands of the people got together, and made a clamor against him, and insisted that he should leave off that design. Some of them also used reproaches, and abused the man, as crowds

of such people usually do. So he habited a great number of his soldiers in their habit, who carried daggers under their garments, and sent them to a place where they might surround them. So he bid the Jews himself go away; but they boldly casting reproaches upon him, he gave the soldiers that signal which had been beforehand agreed on; who laid upon them much greater blows than Pilate had commanded them, and equally punished those that were tumultuous, and those that were not; nor did they spare them in the least: and since the people were unarmed, and were caught by men prepared for what they were about, there were a great number of them slain by this means, and others of them ran away wounded. And thus an end was put to this sedition.

3. Now there was about this time Jesus, a wise man, if it be lawful to call him a man; for he was a doer of wonderful works, a teacher of such men as receive the truth with pleasure. He drew over to him both many of the Jews and many of the Gentiles. He was [the] Christ. And when Pilate, at the suggestion of the principal men amongst us, had condemned him to the cross,[9] those that loved him at the first did not forsake him; for he appeared to them alive again the third day;[10] as the divine prophets had foretold these and ten thousand other wonderful things concerning him. And the tribe of Christians, so named from him, are not extinct at this day.

Publius Cornelius Tacitus

Annals (written c. AD 116), book 15, chapter 44.[1]

This excerpt was written by Publius Cornelius Tacitus, who is widely regarded as one of the greatest Roman historians by modern scholars

Such indeed were the precautions of human wisdom. The next thing was to seek means of propitiating the gods, and recourse was had to the Sibylline books, by the direction of which prayers were offered to Vulcanus, Ceres, and Proserpina. Juno, too, was entreated by the matrons, first, in the Capitol, then on the nearest part of the coast, whence water was procured to sprinkle the fane and image of the goddess. And there were sacred banquets and nightly vigils celebrated by married women. But all human efforts, all the lavish gifts of the emperor, and the propitiations of the gods, did not banish the sinister belief that the conflagration was the result of an order. Consequently, to get rid of the report, Nero fastened the guilt and inflicted the most exquisite tortures on a class hated for their abominations, called Christians by the populace. Christus, from whom the name had its origin, suffered the extreme penalty during the reign of Tiberius at the hands of one of our procurators, Pontius

Pilatus, and a most mischievous superstition, thus checked for the moment, again broke out not only in Judaea, the first source of the evil, but even in Rome, where all things hideous and shameful from every part of the world find their center and become popular. Accordingly, an arrest was first made of all who pleaded guilty; then, upon their information, an immense multitude was convicted, not so much of the crime of firing the city, as of hatred against mankind. Mockery of every sort was added to their deaths. Covered with the skins of beasts, they were torn by dogs and perished, or were nailed to crosses, or were doomed to the flames and burnt, to serve as a nightly illumination, when daylight had expired. Nero offered his gardens for the spectacle, and was exhibiting a show in the circus, while he mingled with the people in the dress of a charioteer or stood aloft on a car. Hence, even for criminals who deserved extreme and exemplary punishment, there arose a feeling of compassion; for it was not, as it seemed, for the public good, but to glut one man's cruelty, that they were being destroyed. *References: Laurentian Library in Florence, Italy.*

Thallus Refers to the Darkness At Christ's Death

About A.D. 52, Thallus wrote a history about the Middle East from the time of the Trojan War to the first century A.D.1 The work has been lost and the only record we have of his writings is through Julius Africanus (AD 221). Below Julius Africanus refers to Christ's crucifixion and the darkness that covered the earth prior to his death.

Thallus Reference to Jesus Christ

"On the whole world there pressed a most fearful darkness; and the rocks were rent by an earthquake, and many places in Judea and other districts were thrown down. This darkness Thallus, in the 263 third book of his history, calls, as appears to me without reason, an eclipse of the sun. For the Hebrews celebrate the passover on the 14th day according to the moon, and the passion of our Savior fails on the day before the passover][see Phlegon]; but an eclipse of the sun takes place only when the moon comes under the sun. And it cannot happen at any other time but in the interval between the first day of the new moon and the last of the old, that is, at their junction: how then should an eclipse be supposed to happen when the moon is almost diametrically opposite the sun?" – Julius Africanus, Chronography, 18.

Conclusion: This reference reveals several key things: Darkness covered the earth at Christ's death. He understood that a solar eclipse could not explain the total darkness.

The time of the darkness agrees with Matthew 27:45.

An eclipse cannot account for the darkness – this was a miracle.

References: 1. F. F. Bruce. The New Testament Documents. W. Eerdmans Publishing Co., 1981.

Chapter 5

The Historicity of the Bible

The chapter explores archaeological and textual evidence supporting the Bible's historical reliability. It begins with the revolutionary discovery of the Dead Sea Scrolls in 1947, explaining how these ancient manuscripts, dating from the third century BC to the first century AD, demonstrate the remarkable accuracy of Biblical transmission.

The chapter examines key archaeological discoveries including the Hammurabi Stele, the Moabite Stone, the Black Obelisk of Shalmaneser III, and the Pontius Pilate Inscription. Each discovery is explained in its historical context, showing how archaeological findings consistently confirm Biblical accounts.

The chapter also discusses the transmission of New Testament texts, examining the more than 5,000 Greek manuscripts and early papyri that authenticate the New Testament's reliability.

Why the Bible is Historically Reliable

In 1947, a discovery was made that became the most important archaeological find of the 20th century. The story begins when a Bedouin shepherd boy named Muhammed was searching for a lost goat. He tossed a stone into a hole in a cliff on the west side of the Dead Sea, about eight miles south of Jericho. To his surprise, he heard the sound of shattering pottery. Investigating, he discovered an amazing sight. On the floor of the cave were several large jars, some of which contained leather scrolls wrapped in linen cloth. Because the jars were carefully sealed, the scrolls had been preserved in excellent condition for nearly nineteen hundred years. They were evidently placed there before the fall of Jerusalem in A.D. 70.

The Value of the Dead Sea Scrolls

Until the discovery of the Qumran scrolls, which date from the third century B.C. to the first century A.D., the oldest Old Testament manuscripts were a fragment of Deuteronomy 6:4 (Nash Papyrus), dated to the first century B.C., a few biblical fragments from the Cairo Geniza (a synagogue storeroom), dating to the fifth century A.D., and the Masoretic texts[1] from the ninth to the 11th centuries A.D.

The oldest existing complete Hebrew manuscript of the Old Testament, the Leningrad Codex, comes from the first decade of the 11th century A.D. The great importance of the Dead Sea Scrolls, therefore, lies in the fact that the earliest scrolls date back to only about two hundred years after the last book of the Old Testament was completed.

Thanks to the Dead Sea Scrolls, we now have a complete manuscript of the Hebrew text of the Book of Isaiah and fragments of most of the other biblical books that are more than one thousand years older than the manuscript previously known to exist.

The significance of this discovery has to do with the detailed closeness of the Isaiah scroll (circa. 125 B.C.) to the Masoretic Text of Isaiah one thousand years later. It demonstrates the unusual accuracy of the copyists of the Scripture over a thousand-year period. When the Masoretic text was compared with the Qumran texts, they were found to be almost identical.

Even though the two copies of Isaiah discovered in Qumran Cave 1 near the Dead Sea in 1947 were a thousand years earlier than the oldest dated manuscript previously known (A.D. 980), they proved to be word-for-word identical with our standard Hebrew Bible in more than 95 percent of the text. The five percent variation consisted chiefly of obvious slips of the pen and variations of spelling. Even those Dead Sea fragments of Deuteronomy and Samuel that point to a different manuscript family from that which underlies our received Hebrew text do not indicate any differences in doctrine or teaching. They do not affect the message of revelation in the slightest.

Thus, we can know that our present Old Testament text, based on the Masoretic text, is practically identical with the Hebrew text in use at the time of Jesus. There is, therefore, no reason to doubt that what the authors of the Old Testament wrote is substantially the same as what we have in our Bibles today.

No other ancient writings comparable to the Old Testament have been transmitted so accurately, mainly because the Jewish scribes and the Masoretes treated God's Word with the greatest imaginable reverence. They devised a complicated system of counting the verses, words, and letters of the text to safeguard against any scribal slips. Any scroll not measuring up to these rules was buried or burned.

The Transmission of the New Testament

All the New Testament books were written during the second half of the first century: Galatians and the two letters to the Thessalonians, around A.D. 50, and John's Gospel and the Book of Revelation, circa. A.D. 90–100.

As with the Old Testament, all the New Testament autographs have been lost. However, because the New Testament books were the most frequently copied and widely circulated books in antiquity, we have today more than five thousand known Greek manuscripts of the New Testament

No other book in antiquity even begins to approach such a large number of extant manuscripts. In comparison, "the Iliad by Homer is second with only 643 manuscripts that still survive. The first complete preserved text of Homer dates from the 13th century."

For Caesar's Gallic War (composed between 58 and 50 B.C.) there are several extant manuscripts, but only nine or 10 are good, and the oldest is some nine hundred years later than Caesar's day. Of the 142 books of the Roman history of Livy (59 B.C.–A.D. 17), only 35 survive; these are known to us from not more than 20 manuscripts of any consequence, only one of which, and that containing fragments of Books II–VI, is as old as the fourth century.

The Manuscripts of the New Testament

The earliest manuscript among the more than five thousand known Greek manuscripts of the New Testament is a small fragment of papyrus (called P52) from around A.D. 130, containing portions, and 38. of John 18:31–33, 37

The Chester Beatty papyri (named after their original owner) come from the second and third centuries and consist of papyri containing portions of all four Gospels and Acts, almost all of Paul's Epistles, the Book of Hebrews, and

Revelation 9 to 17. From the same time period we have the Bodmer papyri (also named after their owner) that contain the Gospels of Luke and John, and the letters to Jude and 1 and 2 Peter. These papyri all come from Egypt, where the dry climate helped to preserve them.

The most complete New Testament manuscripts, written on vellum (parchment), come from the fourth century: (1) Codex Sinaiticus, discovered by Constantine von Tischendorf in St. Catherine's Monastery (at the foot of Mount Sinai), comes from the middle of the fourth century and contains the entire Greek New Testament. (2) Codex Vaticanus, from the Vatican Library, is dated slightly earlier than Sinaiticus and contains the New Testament up to Hebrews 9:14. On textual grounds Codex Vaticanus is considered the most valuable of all existing New Testament manuscripts. Three other important manuscripts are Codex Alexandrinus, Codex Beza, and Codex Ephraemi from the fifth century.

In addition to the approximately 3,200 manuscripts, which are continuous text manuscripts, we have another 2,200 lectionary manuscripts. Lectionaries are "manuscripts in which the text of the New Testament books is divided into separate pericopes [sections], arranged according to their sequence as lessons appointed for the church year."5 Though a few of these lectionaries go back to the fourth century, the majority were written after the eighth century.

New Testament Textual Criticism

We have seen that there is no body of literature in history that enjoys such a wealth of ancient manuscripts as the New Testament. Yet this very fact produces its own problems. The more manuscripts, the greater the textual variations created by scribal mistakes. If a scribe were listening to a dictation, he could make mistakes with words that sound alike; if he were copying from a manuscript before him, he could mistake one word for another that looked like it. Or his eyes could jump from one word to another occurrence of the same word or to another word that had the same ending, and thus a portion of the text could be left out or written twice. Textual critics seek to reconstruct as closely as possible the original wording of the biblical text.

The English classical scholar Sir Frederic Kenyon stated: "It is reassuring at the end to find that the general result of all these discoveries and all this study is to strengthen the proof of the authenticity of the Scriptures, and our conviction that we have in our hands, in substantial integrity, the veritable Word of God." It should also be clearly stated that despite the many variant readings in the manuscripts, none of them affects any point of Christian faith and practice.

Evidence from Archaeology

Though archaeology cannot prove the spiritual truths of the Bible, it can illuminate and clarify the historical circumstances of numerous passages and thereby validate the historicity of many of the events recorded in Scripture. Among the most important discoveries of archaeology that support the historical reliability of Scripture are the following:

1. The Hammurabi Stele (circa. 1700 B.C.) was found by French archaeologists in the winter of 1901–1902 at Susa, the biblical Shushan (Dan. 8:2), and is now exhibited in the Louvre in Paris. It contains about 280 laws, many of which are strikingly similar to the Mosaic laws:

Hammurabi No. 14- If a citizen kidnaps and sells a member of another citizen's household into slavery, then the sentence is death.

Exodus 21:16- And he that stealeth a man, and selleth him, or if he be found in his hand, he shall surely be put to death.

Hammurabi Nos. 196 and 197- If a citizen blinds an eye of an official, then his eye is to be blinded. If one citizen breaks a bone of another, then his own bone is to be broken.

Exodus 21:24- Eye for eye, tooth for tooth, hand for hand, foot for foot,

The discovery of the Hammurabi Stele and other ancient law codes disposed of the old critical view that the laws of the Pentateuch could not have come from the time of Moses.

2. The Merneptah Stele (circa. 1200 B.C.) was found by Sir

Flinders Petrie in the mortuary temple at Thebes and published in 1897. It is today exhibited in Cairo. The stele celebrates Pharaoh Merneptah's (1213-1203 B.C.) victory over rebellious forces in his Asiatic possessions. It contains the earliest reference to the people of Israel in the ancient world.

3. The Moabite Stone (circa. 850 B.C.) is exhibited in the Louvre. In 1868, an Arab sheikh, at Diban, showed the German missionary, F. Klein, an inscribed slab that was 3 feet 10 inches high, 2 feet wide, and 10 inches thick. German and French officials showed interest in the stone. A French orientalist, Ch. Clermont-Ganneau, was able to obtain a "squeeze," i.e., a facsimile impression, of the inscription. This was fortunate because the Arabs, realizing that they had something valuable, broke it into pieces. The fragments were then carried away to bless their grain. Not all the pieces have been recovered, but the inscription has been restored. It recounts the story of the Moabite king Mesha's rebellion against the king of Israel. It supplements the account of Israel's relations with Moab as recorded in 2 Kings 3.

Moabite Stone Omri, ruler of Israel, invaded Moab year after year because Chemosh, the divine patron of Moab, was angry with his people. When the son of Omri succeeded him during my [Moab's ruler's] reign, he bragged: "I too will invade Moab." However, I defeated the son of Omri and drove Israel out of our land forever. Omri and his son ruled the Madaba plains for 40 years.

2 Kings 3:4, 5 – Now Mesha king of Moab was a sheep breeder, and he regularly paid the king of Israel one hundred thousand lambs and the wool of one hundred thousand rams. But it happened, when Ahab died, that the king of Moab rebelled against the king of Israel.

4. The Black Obelisk of Shalmaneser III (circa. 840 B.C.) was discovered in 1846 by A. H. Layard at Nimrud. It is exhibited in the British Museum. It shows the Israelite king Jehu paying tribute to the Assyrian king and provides extrabiblical evidence for the domination of Assyria over Israel as well as the existence of Jehu as king of Israel. "'Also you shall anoint Jehu the son of Nimshi as king over Israel. And Elisha the son of Shaphat of Abel Meholah you shall anoint as prophet in your place'" (1 Kings 19:16, NKJV).7

5. The Taylor Prism (circa. 690 B.C.) is in the British Museum. It was found at Nineveh and contains the military campaigns of Sennacherib (705–681 B.C.), king of Assyria. The best-known passage describes Sennacherib's unsuccessful siege of Jerusalem in the days of Hezekiah, as recorded in 2 Kings 19 and Isaiah 36 and 37. The Assyrian account tacitly agrees with the biblical account by making no claim that Jerusalem was taken. The six-sided hexagonal clay prism says, "I [Hezekiah] made a prisoner in Jerusalem, his royal residence, like a bird in a cage." According to 2 Kings 19:35 and 36,

Sennacherib was unable to capture Jerusalem because "And it came to pass on a certain night that the angel of the Lord went out, and killed in the camp of the Assyrians one hundred and eighty-five thousand. . . . So Sennacherib king of Assyria departed and went away, returned home, and remained at Nineveh."

6. The Tel Dan Stele (ninth or eighth century B.C.) is a black basalt stele erected by an Aramaean king in northernmost Israel, containing an Aramaic inscription to commemorate his victory over the ancient Israelites. Only portions of the inscription remain, but clearly legible is the phrase "house of David" (1 Sam. 20:16). Jehoram, son of Ahab (2 Kings 8:16), also appears in the inscription. This is the first time that the name "David" has been recognized at any archaeological site. Like the Moabite Stone, the Tel Dan Stele seems typical of a memorial intended as a sort of military propaganda, which boasts of Hazael's or his son's victories.

7. The Babylonian Chronicles (sixth century B.C.) are clay tablets that present a concise account of major internal events in Babylonia. They describe the fall of Nineveh in 612 B.C. (Zeph. 2:13, 15), the battle of Carchemish and the submission of Judah, in 605 B.C. (2 Kings 24:7; Dan. 1:2), the capture of Jerusalem in 597 B.C. (2 Kings 24:10–17), and the fall of Babylon to the Persians in 539 B.C. (Isa. 45:1; Dan. 5:30). In connection with the fall of Babylon, the chronicles refer to Belshazzar (Dan. 5:1), who was coregent with his father Nabonidus, the last king of Babylon.

8. The Pontius Pilate Inscription (first century A.D.) was found in 1961 in the theatre of Caesarea Maritima, the city of Pilate's residence in Palestine. Among the few lines still legible are the words "Pontius Pilate Prefect of Judea." The inscription is the first archaeological evidence for Pilate, before whom Jesus was tried and condemned to death (Matt. 27:11–26).

9. Politarch inscriptions Critics of the New Testament claimed that Luke was mistaken in calling the chief magistrates in Thessalonicapolitarchs (Acts 17:6), a title not found in extant classical literature. In the latter half of the 19th century, several inscriptions using this term have been found in Macedonian towns, including Thessalonica.

Apart from these major finds, of which there are quite a few more, there have been many smaller finds, such as rings and seals, that have confirmed the historical reliability of Scripture.

William F. Albright, probably the greatest archaeologist of the 20th century, whose theological position in the 1920s was one of "extreme radicalism," came to appreciate the historical value of Scripture and wrote in 1956, "There can be no doubt that archaeology has confirmed the substantial historicity of the Old Testament tradition."[8] The same is true of the New Testament. Concerning Luke, the historian of the

New Testament, F. F. Bruce wrote, "Our respect for Luke's [historical] reliability continues to grow as our knowledge of this field increases."9

The Evidence from Prophecy

The purpose of prophecy is not to satisfy human curiosity about the future, but to reveal important facts about God's nature—His foreknowledge, His control over all the nations, and His plans for the people of God. In addition, fulfilled prophecies are important evidence for the inspiration and trustworthiness of God's Word. The two prophecies explained below are representative of the many prophecies found in the Old and New Testaments.

Daniel 2 – The Book of Daniel was written in the sixth century B.C., but its prophecies provide evidence for the fact that history is under God's control. Daniel interprets the image in chapter 2 as four successive world empires, beginning with Babylon as the first empire (2:38). The fourth empire would be followed by many smaller kingdoms or nations, symbolized by the 10 toes (vss. 41–43). These nations would continue until God's kingdom, symbolized by the rock "'cut out without hands'" smashing the image to bits (vs. 34), would be established on the earth (vs. 44).

This prophecy found a remarkable fulfilment in history. Babylon was succeeded by three other world empires, Medo-Persia, Greece, and Rome, and Rome was divided up into many smaller kingdoms that still exist in Europe and around the Mediterranean Sea. The only part of the prophecy still unfulfilled is the arrival of the kingdom of God.

Micah 5:2 – According to the prophecy in Micah 5:2, the Messiah would be born in Bethlehem. The Gospels tell us that although the parents of Jesus lived in Nazareth, because of a census in the Roman Empire, Joseph and Mary had to travel to Bethlehem, Joseph's ancestral hometown, where Jesus was born (Luke 2:4–7).

While the Bible is self-authenticating, i.e., the books of Scripture themselves testify to their God-inspired truth, the manuscript evidence, as well as the archaeological and prophetic evidence, confirms the reliability of Scripture. The Dead Sea Scrolls and other manuscript finds have demonstrated the textual reliability of the Bible; and the many archaeological discoveries support the historical reliability of Scripture. Though archaeology cannot prove that the Bible is true, it does confirm the historical background of the Bible. "What biblical archaeology offers to us is the widening of the environment against which we may see the Bible and its world. The canvas is now larger and the context wider."10 Finally, the fulfillment of Biblical prophecies are confirmation of what 2 Peter 1:21 says "prophecy never came by the will of man, but holy men of God spoke as they were moved by the Holy Spirit."

Source: Gerhard Pfandl, Revival and Reformation

Chapter 6

Science In the Bible

This chapter reveals scientific principles found in Scripture that predate their scientific discovery. It examines verses like Isaiah 40:22 (the spherical shape of the earth), Job 26:7 (earth suspended in space), Jeremiah 33:22 (incalculable number of stars), and Ecclesiastes 1:6 (wind circuits).

The chapter explains how these Biblical statements align with modern scientific understanding, including oceanography (Psalm 8:8), meteorology (Job 28:25), and medical science (Leviticus 17:11). Each scientific principle is examined in detail, showing how the Bible accurately described natural phenomena long before their scientific discovery.

The chapter concludes by demonstrating how Genesis 1:1 contains all five of Herbert Spencer's scientific principles: time, force, energy, space, and matter.

The earth is a sphere

Isaiah 40:22 It is he that sitteth upon the circle of the earth, and the inhabitants thereof are as grasshoppers; that stretcheth out the heavens as a curtain, and spreadeth them out as a tent to dwell in.

Incalculable number of stars

Jeremiah 33:22 As the host of heaven cannot be numbered, neither the sand of the sea measured: so will I multiply the seed of David my servant, and the Levites that minister unto me.

Free float of earth in space

Job 26:7 He stretcheth out the north over the empty place, and hangeth the earth upon nothing.

Creation made of invisible elements

Hebrews 11:3 Through faith we understand that the worlds were framed by the word of God, so that things which are seen were not made of things which do appear.

Each star is different

1 Corinthians 15:41 There is one glory of the sun, and another glory of the moon, and another glory of the stars: for one star differeth from another star in glory.

Light moves

Job 38:19-20 Where is the way where light dwelleth? and as for darkness, where is the place thereof.

Air has weight

Job 28:25 To make the weight for the winds; and he weigheth the waters by measure.

Winds blow in cyclones

Ecclesiastes 1:6 The wind goeth toward the south, and turneth about unto the north; it whirleth about continually, and the wind returneth again according to his circuits

Blood is the source of life and health

Leviticus 17:11 For the life of the flesh is in the blood: and I have given it to you upon the altar to make an atonement for your souls: for it is the blood that maketh an atonement for the soul.

Ocean floor contains deep valleys and mountains

2 Samuel 22:16 And the channels of the sea appeared, the foundations of the world were discovered, at the rebuking of the LORD, at the blast of the breath of his nostrils."

Jonah 2:6 I went down to the bottoms of the mountains; the earth with her bars was about me for ever: yet hast thou brought up my life from corruption, O LORD my God.

Oceans contain currents and springs

Psalm 8:8 The fowl of the air, and the fish of the sea, and whatsoever passeth through the paths of the seas.

Job 38:16 Hast thou entered into the springs of the sea? or hast thou walked in the search of the depth?

When dealing with disease, hands should be washed under running water

Leviticus 15:13 And when he that hath an issue is cleansed of his issue; then he shall number to himself seven days for his cleansing, and wash his clothes, and bathe his flesh in running water, and shall be clean.

Genesis 2:6-7 (KJV)

Man = Water + Earth + Air + Fire = Adam = Water + Earth + Air + Fire = Man

+ WATER: But there went up a mist from the earth, and watered the whole face of the ground (Genesis 2:6)

+ EARTH: And the Lord God formed man of the dust of the ground. (Genesis 2:7)

+ AIR: And breathed into his nostrils the breath of life; (Genesis 2:7)

+ FIRE: And man became a living soul. (Genesis 2:7)

Science Proves the Bible

If you've read the Bible, you know that science agrees with Gods word. In fact, science, properly applied and understood, proves the Bible to be inspired. Where the Bible makes a statement relating to a scientific principle or fact, it is completely accurate.

Let us look at some examples of the harmony between science and the Bible. Genesis 1:1 "In the beginning God created the heavens and the earth" This was written by Moses through the inspiration of the Holy Spirit about 1500 B.C. In 1820 A.D. a man named Hubert Spencer gave the world five scientific principles by which man may study the unknown. They are time, force, energy, space, and matter. However, Moses, by inspiration, gave us those scientific principles in Genesis 1:1.

"In the beginning"- time;

"God"- force;

"created"- energy;

"the heavens"- space;

"and the earth"- matter.

All of Spencer's scientific principles are right there in Genesis 1:1.

For many years man has estimated the number of stars in the heavens, and he has increased the estimate many times. Finally, in the 1900's, man determined that the stars could not be counted. God's book has always told us this fact. Notice Genesis 15:5 " And he brought him forth abroad, and said, Look now toward heaven, and tell the stars, if thou be able to number them: and he said unto him, So shall thy seed be."

God was telling Abraham that just as the stars in the heavens cannot be numbered, Abraham's descendants would be more than could be numbered. If man had paid attention to this verse, he would never have tried to count the stars! Another example of how science and the Bible agree relates to the blood in our bodies. Man, now knows that blood is necessary for survival. If our bodies lose too much blood, we will die. However, man did not discover this principle until the 19th century. Before that time, blood-letting was practiced, and many died because draining blood from their bodies drained the very source of life. George Washington, the first President of the United States, is said to have died in this way.

Moses, again writing by inspiration hundreds of years ago, told us something man did not know until much later. " For the life of the flesh is in the blood: and I have given it to you upon the altar to make an atonement for your souls: for it is the blood that maketh an atonement for the soul." (Leviticus 17:11). How could Moses have known about the life-giving qualities of blood unless God had revealed it to him?

In the book of Job, the inspired writer in one verse reveals two scientific principles not known to man until much later. Job. 26:7 "He stretcheth out the north over the empty place, and hangeth the earth upon nothing." There is a place in the North where no stars exist, which cannot be seen with the naked eye. How did the writer of Job know this? Also, the same verse declares that God hangs the earth on nothing. We know this is true, but we have only known it for about 350 years. God's inspired writer told us over 3000 years ago that the earth is held in place by gravitational forces!

When we come to the Psalms, we find an interesting statement in Psalm 8:8. The passage] mentions "The fowl of the air, and the fish of the sea, and whatsoever passeth through the paths of the seas." The phrase, "the paths of the seas" caused a man named Matthew Fontaine Maury to begin a search which led to the discovery of ocean currents, the natural "paths of the seas" created by God. Maury concluded that if God's Book said they were there, they must be there! He was right.

When the Bible makes a statement relating to science, it is always accurate. Notice the Lord's statement for example, in Luke 10:30 Then Jesus answered and said: "And Jesus answering said, *A certain man went down from Jerusalem to Jericho, and fell among thieves, which stripped him of his raiment, and wounded him, and departed, leaving him half dead.*" Now, Jericho is Northeast of Jerusalem, and normally we do not speak of going down when we refer to going North. We generally speak of going down South and up North, don't we? Why did the Lord say the man went down from Jerusalem? It is because Jerusalem is some 2500 feet above sea level. When one leaves Jerusalem in Palestine he goes down to go anywhere in the area. Therefore, our Lord's statement is completely accurate and is recorded accurately by inspiration. No, the Bible is not a geography book, but it is geographically accurate in every instance.

Many archaeologists have explored the land of Palestine. Has any one of those scientists ever discovered anything which disproves the Bible? No. Many archaeological discoveries have confirmed the Biblical record. However, none has ever contradicted the Word of God. This ought to be very reassuring to those of us who believe the Bible to be the inspired Word of God. At the same time, it ought to convince the skeptic, the agnostic, and the atheist, that this Book is God's Book. Therefore, not only does God exist, but He has revealed His Will to man.

Many other examples of the harmony between science and the Bible could be given. However, the ones we have examined are sufficient to show that the Bible is God's Book.

As the apostle Paul declared in 2 Timothy 3:16-17"All scripture is given by inspiration of God, and is profitable for doctrine, for reproof, for correction, for instruction in righteousness: 17 That the man of God may be perfect, thoroughly furnished unto all good works."

The Bible is accurate in matters of science, and it is accurate in the matter of salvation. The Bible tells us that to be saved we must believe in Jesus Christ

In John 8:24 Jesus said "I said therefore unto you, that ye shall die in your sins: for if ye believe not that I am he, ye shall die in your sins" Also in Luke 13:3 the same Lord said, "I tell you, Nay: but, except ye repent, ye shall all likewise perish." In Matthew 10:32-33 Jesus also spoke of confessing Him before men. "32 Whosoever therefore shall confess me before men, him will I confess also before my Father which is in heaven. 33 But whosoever shall deny me before men, him will I also deny before my Father which is in heaven." And in Mark 16:16 Jesus declared, "He that believeth and is baptized shall be saved; but he that believeth not shall be damned." It is then that the Lord Himself adds us to the church of the New Testament, the one body of believers. If the Bible is accurate in matters of science, it is accurate in the matter of salvation.

Source: Kebenaran Bagi Dunia, Truth for the World Ministries.

Chapter 7

The 12 Universal Laws in the Bible

The chapter examines twelve universal laws and their Biblical foundations. Each law is explained in its Biblical context, showing how these universal principles are rooted in scripture.

The Law of Divine Oneness asserts that everything in the universe is interconnected and part of the same whole.

John 17:21 That they all may be one; as thou, Father, *art* in me, and I in thee, that they also may be one in us: that the world may believe that thou hast sent me.

1 Peter 3:8 Finally, be ye all of one mind, having compassion one of another, love as brethren, be pitiful, be courteous:

John 15:5 I am the vine, ye *are* the branches: He that abideth in me, and I him, the same bringeth forth much fruit: for without me ye can do nothing.

John: 14:20 At that day ye shall know that I *am* in my Father, and ye in me, and I in you.

Acts 17:28 Because all people live and move and have their being in God.

Law of Vibration suggests that everything in the universe is in constant motion vibrating at a special frequency.

Hebrews 11:3 Through faith we understand that the worlds were framed by the word of God, so that things which are seen were not made of things which do appear.

Genesis 1:3 And God said, Let there be light: and there was light.

Joshua 6:20 So the people shouted when the priests blew with the trumpets: and it came to pass, when the people heard the sound of the trumpet, and the people shouted with a great shout, that the wall fell down flat, so that the people went up into the city, every man straight before him, and they took the city.

Law of Attraction states that like attracts like.

Proverbs 23:7 For as he thinketh in his heart, so is he.

Matthew 9:29 Then touched he their eyes, saying, According to your faith And be not conformed to this world: but be ye transformed by the renewing of your mind, that ye may prove what is that good, and acceptable, and perfect, will of God.

Matthew 12:35 A good man out of the good treasure of the heart bringeth forth good things: and an evil man out of the evil treasure bringeth forth evil things.

Matthew 21:22 And all things, whatsoever ye shall ask in prayer, believing, ye shall receive.

Law of Correspondence proposes that there are patterns and reflections between the macrocosm (the universe) and the microcosm (the individual).

Matthew 6:10 Your kingdom come, your will be done, on earth as it is in heaven.

Proverbs 4:23 Keep thy heart with all diligence, for out of it are the issues of life.

Romans 12:2 And be not conformed to this world: but be ye transformed by the renewing of your mind, that ye may prove what is that good, and acceptable, and perfect, will of God.

Law of Inspired Action emphasizes the importance of taking inspired action toward one's goals and desires.

James 2:26 For as the body without the spirit is dead, so faith without works is dead also.

Job 22:28 Thou shalt also decree a thing, and it shall be established unto thee:

James 2:18 But someone will say I have faith and you have works show me your faith without your works, and I will show you my faith by my works.

John 14:12 Verily, verily I say unto you, he that believeth on me the works that I do he shall do also; and greater works than these he shall do also

Law of Cause and Effect suggests that every action has a corresponding reaction or consequence. It emphasizes personal responsibility for one's actions and their effects.

Galatians 6:7 Be not deceived; God is not mocked: for whatsoever a man soweth, that shall he also reap.

2 Corinthians 9:6 But this I say, He which soweth sparingly shall reap also sparingly; and he which soweth bountifully shall reap also bountifully.

Matthew 25:29 For unto everyone that hath shall be given, and he shall have abundance: but from him that hath not shall be taken away even that which he hath.

Law of Compensation states that individuals are compensated in accordance with their contributions and the value they bring to the world. It emphasizes the importance of giving and receiving in balance.

2 Corinthians 9:6 But this I say, He which soweth sparingly shall reap also sparingly; and he which soweth bountifully shall reap also bountifully. 7 Every man according as he purposeth in his heart, so let him give; not grudgingly, or of necessity: for God loveth a cheerful giver.

Proverbs 11:24-25 There is that scattereth, and yet increaseth; and there is that withholdeth more than is meet, but it tendeth to poverty. 25 The liberal soul shall be made fat: and he that watereth shall be watered also himself.

Luke 6:38 Give, and it shall be given unto you; good measure, pressed down, and shaken together, and running over, shall men give into your bosom. For with the same measure that ye mete withal it shall be measured to you again.

Law of Perpetual Transmutation of Energy suggests that individuals can transform energy through their thoughts, emotions, and actions.

Romans 12:2 And be not conformed to this world: but be ye transformed by the renewing of your mind, that ye may prove what is that good, and acceptable, and perfect, will of God.

Luke 17:21 Neither shall they say, Lo here! or, lo there! for, behold, the kingdom of God is within you.

Matthew 6:33 But seek ye first the kingdom of God, and his righteousness; and all these things shall be added unto you.

Matthew 19:26 But Jesus beheld them, and said unto them, With men this is impossible; but with God all things are possible.

Hebrews 1:1 Now faith is the substance of things hoped for, and the evidence of things not seen.

Law of Relativity suggests that everything is relative and subjective.

Romans 8:28 And we know that all things work together for good to them that love God, to them who are the called according to his purpose.

Phillipians 4:19 but my God shall supply all your needs according to his riches in glory by Christ Jesus.

Phillipians 4:6-8 Be careful for nothing; but in every thing by prayer and supplication with thanksgiving let your requests be made known unto God.

7 And the peace of God, which passeth all understanding, shall keep your hearts and minds through Christ Jesus.

8 Finally, brethren, whatsoever things are true, whatsoever things are honest, whatsoever things are just, whatsoever things are pure, whatsoever things are lovely, whatsoever things are of good report; if there be any virtue, and if there be any praise, think on these things.

1 Thesselonians 5:18 In every thing give thanks: for this is the will of God in Christ Jesus concerning you.

Law of Polarity states that everything has its opposite—light and dark, hot, and cold, love and fear.

Genesis 1:4 And God saw the light, that it was good: And God divided the light from the darkness.

Isaiah 9:2 The people that walked in darkness have seen a great light: they that dwell in the land of the shadow of death, upon them hath the light shined.

Romans 12:9 Let love be without dissimulation. Abhor that which is evil; cleave to that which is good.

Law of Rhythm suggests that everything in the universe operates in cycles and rhythms.

Ecclesiastes 3:1-8 To every thing there is a season, and a time to every purpose under the heaven:

2 A time to be born, and a time to die; a time to plant, and a time to pluck up that which is planted;

3 A time to kill, and a time to heal; a time to break down, and a time to build up;

4 A time to weep, and a time to laugh; a time to mourn, and a time to dance;

5 A time to cast away stones, and a time to gather stones together; a time to embrace, and a time to refrain from embracing;

6 A time to get, and a time to lose; a time to keep, and a time to cast away;

7 A time to rend, and a time to sew; a time to keep silence, and a time to speak;

8 A time to love, and a time to hate; a time of war, and a time of peace.

Law of Gender suggests that everything has masculine and feminine aspects, which are present in all creation.

Genesis 1:27 So God created man in his own image, in the image of God he created him; In the image of God he created them.

1 Corinthians 11:8-9 For the man is not of the woman: but the woman of the man. Neither was the man created for the woman; but the woman for the man.

Matthew 19:4 "And he answered and said unto them, have ye not read, that he which made them at the beginning made them male and female.

Chapter 8

Salvation

This chapter presents the Biblical doctrine of salvation through several key themes. It begins with the fundamental principle that salvation comes through faith in Jesus Christ (Romans 10:13), explaining the concept of 'once saved, always saved' (John 10:28-29).

The chapter explores topics including divine forgiveness (Ephesians 4:31-32), faith (Matthew 17:20), authority in Jesus (Luke 10:19), trust (Jeremiah 29:11), love (1 Corinthians 13), and gratitude (1 Chronicles 16:34).

Special attention is given to the concept that believers are already seated in heavenly places (Ephesians 2:6), explaining how this spiritual reality affects daily life. The chapter concludes with an examination of spiritual gifts and their role in the believer's life.

For whosoever shall call upon the name of the Lord shall be saved (KJV)

Romans 10:11-15 For with the heart man believeth unto righteousness; and with the mouth confession is made unto salvation. 11 For the scripture saith, Whosoever believeth on him shall not be ashamed. 12 For there is no difference between the Jew and the Greek: for the same Lord over all is rich unto all that call upon him. 13 For whosoever shall call upon the name of the Lord shall be saved. 14 How then shall they call on him in whom they have not believed? and how shall they believe in him of whom they have not heard? and how shall they hear without a preacher? 15 And how shall they preach, except they be sent? as it is written, How beautiful are the feet of them that preach the gospel of peace, and bring glad tidings of good things! 16But they have not all obeyed the gospel. For Esaias saith, Lord, who hath believed our report?

Once Saved Always Saved

John 10:25-31 (KJV)

25 Jesus answered them, I told you, and ye believed not: the works that I do in my Father's name, they bear witness of me.

26 But ye believe not, because ye are not of my sheep, as I said unto you.

27 My sheep hear my voice, and I know them, and they follow me:

28 And I give unto them eternal life; and they shall never perish, neither shall any man pluck them out of my hand.

29 My Father, which gave them me, is greater than all; and no man is able to pluck them out of my Father's hand.

30 I and my Father are one.

31 "Then the Jews took up stones again to stone him.

God Doesn't Remember Our Sins (KJV)

Hebrews 8: 9-13 Not according to the covenant that I made with their fathers in the day when I took them by the hand to lead them out of the land of Egypt; because they continued not in my covenant, and I regarded them not, saith the Lord.

10 For this is the covenant that I will make with the house of Israel after those days, saith the Lord; I will put my laws into their mind, and write them in their hearts: and I will be to them a God, and they shall be to me a people:

11 And they shall not teach every man his neighbour, and every man his brother, saying, Know the Lord: for all shall know me, from the least to the greatest.

12 For I will be merciful to their unrighteousness, and their sins and their iniquities will I remember no more.

13 In that he saith, A new covenant, he hath made the first old. Now that which decayeth and waxeth old is ready to vanish away.

We in Christ are Free from the Law (KJV)

Romans 10:4 For Christ is the end of the law for righteousness to every one that believeth.

Galatians 3:10-14 For as many as are of the works of the law are under the curse: for it is written, Cursed is every one that continueth not in all things which are written in the book of the law to do them.

11 But that no man is justified by the law in the sight of God, it is evident: for, The just shall live by faith.

12 And the law is not of faith: but, The man that doeth them shall live in them.

13 Christ hath redeemed us from the curse of the law, being made a curse for us: for it is written, Cursed is every one that hangeth on a tree:

Romans 2:12 For as many as have sinned without law shall also perish without law: and as many as have sinned in the law shall be judged by the law.

Galatians 5:18 But if ye be led of the Spirit, ye are not under the law.

Romans 3: 25-31 Whom God hath set forth to be a

propitiation through faith in his blood, to declare his righteousness for the remission of sins that are past, through the forbearance of God;

26 To declare, I say, at this time his righteousness: that he might be just, and the justifier of him which believeth in Jesus.

27 Where is boasting then? It is excluded. By what law? of works? Nay: but by the law of faith.

28 **Therefore we conclude that a man is justified by faith without the deeds of the law**.

29 Is he the God of the Jews only? is he not also of the Gentiles? Yes, of the Gentiles also:

30 Seeing it is one God, which shall justify the circumcision by faith, and uncircumcision through faith.

31 Do we then make void the law through faith? God forbid: yea, we establish the law.

Forgiveness (KJV)

Romans 19 Dearly beloved, avenge not yourselves, but rather give place unto wrath: for it is written, Vengeance is mine; I will repay, saith the Lord.

Ephesians 4:31-32 Let all bitterness, and wrath, and anger, and clamour, and evil speaking, be put away from you, with all malice: 32 And be ye kind one to another, tenderhearted, forgiving one another, even as God for Christ's sake hath forgiven you.

Mark 11:25 And when ye stand praying, forgive, if ye have ought against any: that your Father also which is in heaven may forgive you your trespasses.

Matthew 6:15 But if ye forgive not men their trespasses, neither will your Father forgive your trespasses.

Luke 6:37 Judge not, and ye shall not be judged: condemn not, and ye shall not be condemned: forgive, and ye shall be forgiven:

Romans 12:19 Dearly beloved, avenge not yourselves, but rather give place unto wrath: for it is written, Vengeance is mine; I will repay, saith the Lord.

Matthew 18:21-22 Then came Peter to him, and said, Lord, how oft shall my brother sin against me, and I forgive him? till seven times?

22 Jesus saith unto him, I say not unto thee, Until seven times: but, Until seventy times seven.

Colossians 3:13 Forbearing one another, and forgiving one another, if any man have a quarrel against any: even as Christ forgave you, so also do ye.

Ephesians 4:31-32 Let all bitterness, and wrath, and anger, and clamour, and evil speaking, be put away from you, with all malice:

32 And be ye kind one to another, tenderhearted, forgiving one another, even as God for Christ's sake hath forgiven you.

Faith (KJV)

17 Then Jesus answered and said, O faithless and perverse generation, how long shall I be with you? how long shall I suffer you? bring him hither to me.

18 And Jesus rebuked the devil; and he departed out of him: and the child was cured from that very hour.

19 Then came the disciples to Jesus apart, and said, Why could not we cast him out?

20 And Jesus said unto them, Because of your unbelief: for verily I say unto you, If ye have faith as a grain of mustard seed, ye shall say unto this mountain, Remove hence to yonder place; and it shall remove; and nothing shall be impossible unto you.

21 Howbeit this kind goeth not out but by prayer and fasting.

22 And while they abode in Galilee, Jesus said unto them, "The Son of man shall be betrayed into the hands of men:

23 And they shall kill him, and the third day he shall be raised again. And they were exceeding sorry.

Our Authority in Jesus (KJV)

Genesis 1:26 And God said, Let us make man in our image, after our likeness: and let them have dominion over the fish of the sea, and over the fowl of the air, and over the cattle, and over all the earth, and over every creeping thing that creepeth upon the earth.

Ephesians 2:5-6 Even when we were dead in sins, hath quickened us together with Christ, by grace ye are saved

6 And hath raised us up together, and made us sit together in heavenly places in Christ Jesus

Luke 10:19 Behold, I give unto you power to tread on serpents and scorpions, and over all the power of the enemy: and nothing shall by any means hurt you.

Matthew 28:18 And Jesus came and spake unto them, saying, All power is given unto me in heaven and in earth.

John 1:12 But as many as received him, to them gave he power to become the sons of God, even to them that believe on his name:

Matthew 10:8 Heal the sick, cleanse the lepers, raise the dead, cast out devils: freely ye have received, freely give.

Trust in Jesus (KJV)

Jeremiah 29:11 For I know the thoughts that I think toward you, saith the LORD, thoughts of peace, and not of evil, to give you an expected end.

Proverbs 3:5-6 Trust in the Lord with all thine heart; and lean not unto thine own understanding.

6 In all thy ways acknowledge him, and he shall direct thy paths

Psalm 25:2 O my God, I trust in thee: let me not be ashamed, let not mine enemies triumph over me.

Psalm 91:2 I will say of the Lord, He is my refuge and my fortress: my God; in him will I trust.

Isaiah 26:3-4 Thou wilt keep him in perfect peace, whose mind is stayed on thee: because he trusteth in thee.

4 Trust ye in the LORD for ever: for in the LORD JEHOVAH is everlasting strength:

Psalms 37:5 Commit thy way unto the LORD; trust also in him; and he shall bring it to pass.

Psalms 31:14-15 But I trusted in thee, O LORD: I said, Thou art my God.

15 My times are in thy hand: deliver me from the hand of mine enemies, and from them that persecute me.

Love (KJV)

Song of Solomon 8:7 Many waters cannot quench love, neither can the floods drown it: if a man would give all the substance of his house for love, it would utterly be contemned.

Matthew 22:37-39 Jesus said unto him, Thou shalt love the Lord thy God with all thy heart, and with all thy soul, and with all thy mind. This is the first and great commandment. And the second is like unto it, Thou shalt love thy neighbour as thyself.

John 3:16 For God so loved the world, that he gave his only begotten Son, that whosoever believeth in him should not perish, but have everlasting life.

John 13:34-35 A new commandment I give unto you, That ye love one another; as I have loved you, that ye also love one another. By this shall all men know that ye are my disciples, if ye have love one to another.

Romans 8:38-39 For I am persuaded, that neither death, nor life, nor angels, nor principalities, nor powers, nor things present, nor things to come, Nor height, nor depth, nor any other creature, shall be able to separate us from the love of God, which is in Christ Jesus our Lord.

Romans 13:8 Owe no man any thing, but to love one

another: for he that loveth another hath fulfilled the law.

1 Corinthians 13:4-7 Charity suffereth long, and is kind; charity envieth not; charity vaunteth not itself, is not puffed up, Doth not behave itself unseemly, seeketh not her own, is not easily provoked, thinketh no evil; Rejoiceth not in iniquity, but rejoiceth in the truth; Beareth all things, believeth all things, hopeth all things, endureth all things.

1 Corinthians 13:13 And now abideth faith, hope, charity, these three; but the greatest of these is charity.

1 John 4:7-8 Beloved, let us love one another: for love is of God; and every one that loveth is born of God, and knoweth God. He that loveth not knoweth not God; for God is love.

1 John 4:18 There is no fear in love; but perfect love casteth out fear: because fear hath torment. He that feareth is not made perfect in love.

Gratitude (KJV)

1 Chronicles 16:8 Give thanks unto the Lord, call upon his name, make known his deeds among the people.

1 Timothy 4:4 For every creature of God is good, and nothing to be refused, if it be received with thanksgiving:

1 Chronicles 16:35 And say ye, Save us, O God of our salvation, and gather us together, and deliver us from the heathen, that we may give thanks to thy holy name, and glory in thy praise.

1 Chronicles 16:34 O give thanks unto the Lord; for he is good; for his mercy endureth for ever.

Psalm 18:49 Therefore will I give thanks unto thee, O Lord, among the heathen, and sing praises unto thy name.

Psalm 30:4 Sing unto the Lord, O ye saints of his, and give thanks at the remembrance of his holiness.

Psalm 30:12 To the end that my glory may sing praise to thee, and not be silent. O Lord my God, I will give thanks unto thee for ever.

1 Corinthians 15:57 But thanks be to God, which giveth us the victory through our Lord Jesus Christ.

2 Corinthians 4:15 For all things are for your sakes, that the abundant grace might through the thanksgiving of many redound to the glory of God.

2 Corinthians 9:11 Being enriched in every thing to all bountifulness, which causeth through us thanksgiving to God.

Colossians 3:17 And whatsoever ye do in word or deed, do all in the name of the Lord Jesus, giving thanks to God and the Father by him.

Isaiah 12:4-5 And in that day shall ye say, Praise the Lord, call upon his name, declare his doings among the people, make mention that his name is exalted.

5 Sing unto the Lord; for he hath done excellent things: this is known in all the earth.

The Kingdom of God is Within

Meditation and Prayer (KJV)

Matthew 6:6 But thou, when thou prayest, enter into thy closet, and when thou hast shut thy door, pray to thy Father which is in secret; and thy Father which seeth in secret shall reward thee openly.

Psalm 63:6 When I remember thee upon my bed, and meditate on thee in the night watches.

Psalm 19:14 Let the words of my mouth, and the meditation of my heart, be acceptable in thy sight, O LORD, my strength, and my redeemer.

Psalm 119:15 I will meditate in thy precepts and, have respect unto thy ways.

Psalm 104:34 My meditation of him shall be sweet: I will be glad in the LORD.

Psalm 49:3 My mouth shall speak of wisdom; and the meditation of my heart shall be of understanding.

Isaiah 26:3 Thou wilt keep him in perfect peace, whose mind is stayed on thee:

because he trusteth in thee.

Psalm 77:12 I will meditate also of all thy work, and talk of thy doings.

1 Timothy :15 Meditate upon these things; give thyself wholly to them; that thy profiting may appear to all.

Philippians 4:8 Finally, brethren, whatsoever things are true, whatsoever things are honest, whatsoever things are just, whatsoever things are pure, whatsoever things are lovely, whatsoever things are of good report; if there be any virtue, and if there be any praise, think on these things.

Spiritual Gifts	
Discernment	Wisdom
Knowledge	Prophecy
Healing	Faith
Miracles	Exhortation
Giving	Leadership
Mercy	Teaching
Service	Speaking/Interpretation of Tongues

We Are in Heaven Now (KJV)

We who are saved by grace through faith in Jesus, are already in heaven. Through prayer and meditation on the good report we can experience the kingdom of God within.

Luke 17:21 Neither shall they say, Lo here! or, lo there! for, behold**, the kingdom of God is within you.**

2 Corinthians 5:8 We are confident, I say, and willing rather to be absent from the body, and to be **present with the Lord.**

Colossians 3:1-3 If ye then be risen with Christ, seek those things which are above, where Christ sitteth on the right hand of God.

2 Set your affection on things above, not on things on the earth.

3 For ye are dead, and your life is hid with Christ in God.

Ephesians 2:6-7 And hath raised us up together, and made us sit together in heavenly places in Christ Jesus:

7 That in the ages to come he might shew the exceeding riches of his grace in his kindness toward us through Christ Jesus.

Matthew 6:33 But seek first the kingdom of God and His righteousness, and all these things shall be added to you.

Romans 8:9 But you are not in the flesh but in the Spirit, if indeed the Spirit of God dwells in you. Now if anyone does not have the Spirit of Christ, he is not His.

Hebrews 11:9-10 By faith he sojourned in the land of promise, as in a strange country, dwelling in tabernacles with Isaac and Jacob, the heirs with him of the same promise:

10 For he looked for a city which hath foundations, whose builder and maker is God.

Chapter 9

Psalms

The final chapter presents a collection of key Psalms that express the full range of human experience with God. It includes Psalm 23 (the shepherd's psalm), Psalm 91 (divine protection), Psalm 121 (God's constant care), Psalm 34 (praise and deliverance), Psalm 100 (joyful worship), and others.

Each Psalm is presented in its entirety, with its themes of praise, thanksgiving, trust, and divine protection highlighted.

The chapter shows how these ancient prayers and songs remain relevant for modern readers, offering comfort, guidance, and inspiration.

Psalm 77 (KJV)

1 I cried unto God with my voice, even unto God with my voice; and he gave ear unto me.

2 In the day of my trouble I sought the Lord: my sore ran in the night, and ceased not: my soul refused to be comforted.

3 I remembered God, and was troubled: I complained, and my spirit was overwhelmed. Selah.

4 Thou holdest mine eyes waking: I am so troubled that I cannot speak.

5 I have considered the days of old, the years of ancient times.

6 I call to remembrance my song in the night: I commune with mine own heart: and my spirit made diligent search.

7 Will the Lord cast off for ever? and will he be favourable no more?

8 Is his mercy clean gone for ever? doth his promise fail for evermore?

9 Hath God forgotten to be gracious? hath he in anger shut up his tender mercies? Selah.

10 And I said, This is my infirmity: but I will remember the years of the right hand of the most High.

11 I will remember the works of the Lord: surely I will remember thy wonders of old.

12 I will meditate also of all thy work, and talk of thy doings.

13 Thy way, O God, is in the sanctuary: who is so great a God as our God?

14 Thou art the God that doest wonders: thou hast declared thy strength among the people.

15 Thou hast with thine arm redeemed thy people, the sons of Jacob and Joseph. Selah.

16 The waters saw thee, O God, the waters saw thee; they were afraid: the depths also were troubled.

17 The clouds poured out water: the skies sent out a sound: thine arrows also went abroad.

18 The voice of thy thunder was in the heaven: the lightnings lightened the world: the earth trembled and shook.

19 Thy way is in the sea, and thy path in the great waters, and thy footsteps are not known.

20 Thou leddest thy people like a flock by the hand of Moses and Aaron.

Psalm 23 (KJV)

1 23 The Lord is my shepherd; I shall not want.

2 He maketh me to lie down in green pastures: he leadeth me beside the

still waters.

3 He restoreth my soul: he leadeth me in the paths of righteousness for his name's sake.

4 Yea, though I walk through the valley of the shadow of death, I will fear no evil: for thou art with me; thy rod and thy staff they comfort me.

5 Thou preparest a table before me in the presence of mine enemies: thou anointest my head with oil; my cup runneth over.

6 Surely goodness and mercy shall follow me all the days of my life: and I will dwell in the house of the Lord for ever.

Psalm 91 (KJV)

1 He that dwelleth in the secret place of the most High shall abide under the shadow of the Almighty.

2 I will say of the Lord, He is my refuge and my fortress: my God; in him will I trust.

3 Surely he shall deliver thee from the snare of the fowler, and from the noisome pestilence.

4 He shall cover thee with his feathers, and under his wings shalt thou trust: his truth shall be thy shield and buckler.

5 Thou shalt not be afraid for the terror by night; nor for the arrow that flieth by day;

6 Nor for the pestilence that walketh in darkness; nor for the destruction that wasteth at noonday.

7 A thousand shall fall at thy side, and ten thousand at thy right hand; but it shall not come nigh thee.

8 Only with thine eyes shalt thou behold and see the reward of the wicked.

9 Because thou hast made the Lord, which is my refuge, even the most High, thy habitation;

10 There shall no evil befall thee, neither shall any plague come nigh thy dwelling.

11 For he shall give his angels charge over thee, to keep thee in all thy ways.

12 They shall bear thee up in their hands, lest thou dash thy foot against a stone.

13 Thou shalt tread upon the lion and adder: the young lion and the dragon shalt thou trample under feet.

14 Because he hath set his love upon me, therefore will I deliver him: I will set him on high, because he hath known my name.

15 He shall call upon me, and I will answer him: I will be with him in trouble; I will deliver him, and honour him.

16 With long life will I satisfy him, and shew him my salvation.

Psalm 121 (KJV)

1 I will lift up mine eyes unto the hills, from whence cometh my help.

2 My help cometh from the Lord, which made heaven and earth.

3 He will not suffer thy foot to be moved: he that keepeth thee will not slumber.

4 Behold, he that keepeth Israel shall neither slumber nor sleep.

5 The Lord is thy keeper: the Lord is thy shade upon thy right hand.

6 The sun shall not smite thee by day, nor the moon by night.

7 The Lord shall preserve thee from all evil: he shall preserve thy soul.

8 The Lord shall preserve thy going out and thy coming in from this time forth, and even for evermore.

Psalm 59 (KJV)

1 Deliver me from mine enemies, O my God: defend me from them that rise up against me.

2 Deliver me from the workers of iniquity, and save me from bloody men.

3 For, lo, they lie in wait for my soul: the mighty are gathered against me; not for my transgression, nor for my sin, O Lord.

4 They run and prepare themselves without my fault: awake to help me, and behold.

5 Thou therefore, O Lord God of hosts, the God of Israel, awake to visit all the heathen: be not merciful to any wicked transgressors. Selah.

6 They return at evening: they make a noise like a dog, and go round about the city.

7 Behold, they belch out with their mouth: swords are in their lips: for who, say they, doth hear?

8 But thou, O Lord, shalt laugh at them; thou shalt have all the heathen in derision.

9 Because of his strength will I wait upon thee: for God is my

defence.

10 The God of my mercy shall prevent me: God shall let me see my desire upon mine enemies.

11 Slay them not, lest my people forget: scatter them by thy power; and bring them down, O Lord our shield.

12 For the sin of their mouth and the words of their lips let them even be taken in their pride: and for cursing and lying which they speak.

13 Consume them in wrath, consume them, that they may not be: and let them know that God ruleth in Jacob unto the ends of the earth. Selah.

14 And at evening let them return; and let them make a noise like a dog, and go round about the city.

15 Let them wander up and down for meat, and grudge if they be not satisfied.

16 But I will sing of thy power; yea, I will sing aloud of thy mercy in the morning: for thou hast been my defence and refuge in the day of my trouble.

17 Unto thee, O my strength, will I sing: for God is my defence, and the God of my mercy.

Psalm 146 (KJV)

1 Praise ye the Lord. Praise the Lord, O my soul.

2 While I live will I praise the Lord: I will sing praises unto my God while I have any being.

3 Put not your trust in princes, nor in the son of man, in whom there is no help.

4 His breath goeth forth, he returneth to his earth; in that very day his thoughts perish.

5 Happy is he that hath the God of Jacob for his help, whose hope is in the Lord his God:

6 Which made heaven, and earth, the sea, and all that therein is: which keepeth truth for ever:

7 Which executeth judgment for the oppressed: which giveth food to the hungry. The Lord looseth the prisoners:

8 The Lord openeth the eyes of the blind: the Lord raiseth them that are bowed down: the Lord loveth the righteous:

9 The Lord preserveth the strangers; he relieveth the fatherless and widow: but the way of the wicked he turneth upside down.

10 The Lord shall reign for ever, even thy God, O Zion, unto all generations. Praise ye the Lord.

Psalm 34 (KJV)

1 I will bless the LORD at all times: his praise shall continually be in my mouth.

2 My soul shall make her boast in the LORD: the humble shall hear thereof, and be glad.

3 O magnify the LORD with me, and let us exalt his name together.

4 I sought the LORD, and he heard me, and delivered me from all my fears.

5 They looked unto him, and were lightened: and their faces were not ashamed.

6 This poor man cried, and the LORD heard him, and saved him out of all his troubles.

7 The angel of the LORD encampeth round about them that fear him, and delivereth them.

8 O taste and see that the LORD is good: blessed is the man that trusteth in him.

9 O fear the LORD, ye his saints: for there is no want to them that fear him.

10 The young lions do lack, and suffer hunger: but they that seek the LORD shall not want any good thing.

11 Come, ye children, hearken unto me: I will teach you the fear of the LORD.

12 What man is he that desireth life, and loveth many days, that he may see good?

13 Keep thy tongue from evil, and thy lips from speaking guile.

14 Depart from evil, and do good; seek peace, and pursue it.

15 The eyes of the LORD are upon the righteous, and his ears are open unto their cry.

16 The face of the LORD is against them that do evil, to cut off the remembrance of them from the earth.

17 The righteous cry, and the LORD heareth, and delivereth them out of all their troubles.

18 The LORD is nigh unto them that are of a broken heart; and saveth such as be of a contrite spirit.

19 Many are the afflictions of the righteous: but the LORD delivereth him out of them all.

20 He keepeth all his bones: not one of them is broken.

21 Evil shall slay the wicked: and they that hate the righteous shall be desolate.

22 The LORD redeemeth the soul of his servants: and none of them that trust in him shall be desolate.

Psalm 100 (KJV)

1 Make a joyful noise unto the LORD, all ye lands.

2 Serve the LORD with gladness: come before his presence with singing.

3 Know ye that the LORD he is God: it is he that hath made us, and not we ourselves; we are his people, and the sheep of his pasture. 4 Enter into his gates with thanksgiving, and into his courts with praise: be thankful unto him, and bless his name.

5 For the LORD is good; his mercy is everlasting; and his truth endureth to all generations.

Psalm 92 (KJV)

1 It is a good thing to give thanks unto the LORD, and to sing praises unto thy name, O most High:

2 To shew forth thy lovingkindness in the morning, and thy faithfulness every night,

3 Upon an instrument of ten strings, and upon the psaltery; upon the harp with a solemn sound.

4 For thou, LORD, hast made me glad through thy work: I will triumph in the works of thy hands.

5 O LORD, how great are thy works! and thy thoughts are very deep.

6 A brutish man knoweth not; neither doth a fool understand this.

7 When the wicked spring as the grass, and when all the workers of iniquity do flourish; it is that they shall be destroyed for ever:

8 But thou, LORD, art most high for evermore.

9 For, lo, thine enemies, O LORD, for, lo, thine enemies shall perish; all the workers of iniquity shall be scattered.

10 But my horn shalt thou exalt like the horn of an unicorn: I shall be anointed with fresh oil.

11 Mine eye also shall see my desire on mine enemies, and mine ears shall hear my desire of the wicked that rise up against me.

12 The righteous shall flourish like the palm tree: he shall grow like a cedar in Lebanon.

13 Those that be planted in the house of the LORD shall flourish in the courts of our God.

14 They shall still bring forth fruit in old age; they shall be fat and flourishing;

15 To shew that the LORD is upright: he is my rock, and there is no unrighteousness* in him

Psalm 136 (KJV)

1 O give thanks unto the Lord; for he is good: for his mercy endureth for ever.

2 O give thanks unto the God of gods: for his mercy endureth for ever.

3 O give thanks to the Lord of lords: for his mercy endureth for ever.

4 To him who alone doeth great wonders: for his mercy endureth for ever.

5 To him that by wisdom made the heavens: for his mercy endureth for ever.

6 To him that stretched out the earth above the waters: for his mercy endureth for ever.

7 To him that made great lights: for his mercy endureth for ever:

8 The sun to rule by day: for his mercy endureth for ever:

9 The moon and stars to rule by night: for his mercy endureth for ever.

10 To him that smote Egypt in their firstborn: for his mercy

endureth for ever:

11 And brought out Israel from among them: for his mercy endureth for ever:

12 With a strong hand, and with a stretched out arm: for his mercy endureth for ever.

13 To him which divided the Red sea into parts: for his mercy endureth for ever:

14 And made Israel to pass through the midst of it: for his mercy endureth for ever:

15 But overthrew Pharaoh and his host in the Red sea: for his mercy endureth for ever.

16 To him which led his people through the wilderness: for his mercy endureth for ever.

17 To him which smote great kings: for his mercy endureth for ever:

18 And slew famous kings: for his mercy endureth for ever:

19 Sihon king of the Amorites: for his mercy endureth for ever:

20 And Og the king of Bashan: for his mercy endureth for ever:

21 And gave their land for an heritage: for his mercy endureth for ever:

22 Even an heritage unto Israel his servant: for his mercy endureth for ever.

23 Who remembered us in our low estate: for his mercy endureth for ever:

24 And hath redeemed us from our enemies: for his mercy endureth for ever.

25 Who giveth food to all flesh: for his mercy endureth for ever.

26 O give thanks unto the God of heaven: for his mercy endureth for ever.

Psalm 3 (KJV)

1 Lord, how are they increased that trouble me! many are they that rise up against me.

2 Many there be which say of my soul, There is no help for him in God. Selah.

3 But thou, O Lord, art a shield for me; my glory, and the lifter up of mine head.

4 I cried unto the Lord with my voice, and he heard me out of his holy hill. Selah.

5 I laid me down and slept; I awaked; for the Lord sustained me.

6 I will not be afraid of ten thousands of people, that have set themselves against me round about.

7 Arise, O Lord; save me, O my God: for thou hast smitten all mine enemies upon the cheek bone; thou hast broken the teeth of the ungodly.

8 Salvation belongeth unto the Lord: thy blessing is upon thy people. Selah.

Psalm 17 (KJV)

1 Hear the right, O Lord, attend unto my cry, give ear unto my prayer, that goeth not out of feigned lips.

2 Let my sentence come forth from thy presence; let thine eyes behold the things that are equal.

3 Thou hast proved mine heart; thou hast visited me in the night; thou hast tried me, and shalt find nothing; I am purposed that my mouth shall not transgress.

4 Concerning the works of men, by the word of thy lips I have kept me from the paths of the destroyer.

5 Hold up my goings in thy paths, that my footsteps slip not.

6 I have called upon thee, for thou wilt hear me, O God: incline thine ear unto me, and hear my speech.

7 Shew thy marvellous lovingkindness, O thou that savest by thy right hand them which put their trust in thee from those that rise up against them.

8 Keep me as the apple of the eye, hide me under the shadow of thy wings,

9 From the wicked that oppress me, from my deadly enemies, who compass me about.

10 They are inclosed in their own fat: with their mouth they speak proudly.

11 They have now compassed us in our steps: they have set their eyes bowing down to the earth;

12 Like as a lion that is greedy of his prey, and as it were a young lion lurking in secret places.

13 Arise, O Lord, disappoint him, cast him down: deliver my soul from the wicked, which is thy sword:

14 From men which are thy hand, O Lord, from men of the world, which have their portion in this life, and whose belly thou fillest with thy hid treasure: they are full of children, and leave the rest of their substance to their babes.

15 As for me, I will behold thy face in righteousness: I shall be satisfied, when I awake, with thy likeness.

Afterword

As we draw this journey to a close, we hope that you have discovered and embraced Jesus Christ in a profound and transformative way. Our goal is that you know Him as God incarnate—Emmanuel, God with us—the pre-existent and eternal Word who took on flesh in the person of Jesus and chose to live among us in tender proximity and humble service.

Through the course of these pages, we have delved deeply into compelling evidence spanning the fascinating realms of mathematics, archaeology, prophecy, and history. Together, we have traced the Fibonacci sequence spiraling not only through the vastness of galaxies but also through the intricate blossom of every flowering plant. We've uncovered ancient manuscripts meticulously preserved through the ages, boldly confirming biblical narratives. We have marveled at the astronomical improbabilities of fulfilled prophecies and calculated their overwhelming odds. Yet, as wondrous as these intellectual explorations are, and as substantial as the facts may be, they all point to something immeasurably more profound than mere intellectual acknowledgment or academic assent.

They point directly to the very heart and essence of God's nature. For God is love, and gratitude, and the Kingdom of God is within.

This Creator, who wove His signature with intricate precision into the DNA of every living cell and scattered it as a mark across the entire cosmos, is ever present. The same creative mind that calculated and fine-tuned the precise gravitational constants necessary for stars to form and for life to flourish—that very same mind loves us personally, intimately, and with unimaginable passion.

This relationship transcends mere intellectual assent. When Scripture speaks of being "in Christ," it describes our truest state of being. We have died to our old identities, our old ways of perceiving reality. In Christ, we participate in heaven now—not as a distant realm we hope to reach someday, but as a present reality within, here and now. We are not of the world, for our true citizenship lies in the Kingdom of Heaven.

** All scripture used in Harmony were taken from the King James Version of the Bible.

*Thank You
Jesus
for All
the Best Things*

www.ingramcontent.com/pod-product-compliance
Lightning Source LLC
Chambersburg PA
CBHW050340010526
44119CB00049B/634